2

WHALING AND OLD SALEM

(A Chronicle of the Sea)

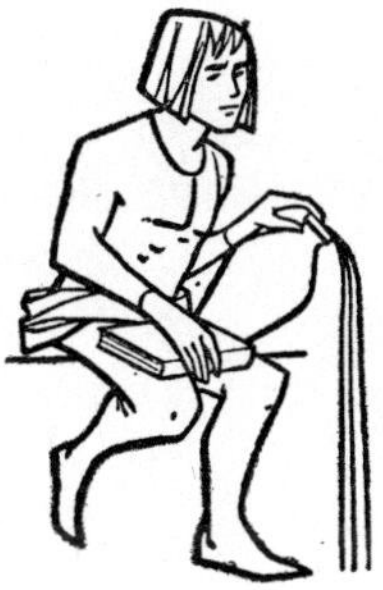

Also by FRANCES DIANE ROBOTTI

CHRONICLES OF OLD SALEM *(an epic of New England)*
MUCH DEPENDS ON DINNER *(the table cloth game)*

Whaling

AND

Old Salem

(A CHRONICLE OF THE SEA)

by

Frances Diane Robotti

With an Introduction by

HENRY BEETLE HOUGH

Fountainhead Publishers

475 FIFTH AVENUE, NEW YORK 17, N. Y.

First Printing

Library of Congress Catalog Card No. 61-14824

Published in New York by Fountainhead Publishers, Inc. and simultaneously in Toronto, Canada, by Ambassador Books, Ltd.

Manufactured in the United States of America by The Colonial Press Inc., Clinton, Mass.

This book is dedicated to

JAMES W. HARBIN, JR.

of the Nautical Research Guild
for his encouragement to me—
a Guild distaffer—in preserving
the whaleman's achievement.

Author's Preface and Acknowledgments

As old-time whaling of harpoon and lance recedes farther into the background of passing time, interest remains high in the most picturesque and daring of America's lost industries. Salem, Massachusetts, in particular, led in trading with exotic places of the world as New England seaports created their legendary tradition of geographical exploration and discovery during their quest of new markets and products. Salem captains and merchants effected in the Orient, South America and on the continent a mercantile exchange which was early translated into the opulence of gracious living, the enjoyment of elegant clothing, exotic foods and furnishings against a background of the architectural purity of Samuel McIntyre mansions along the elm-shaded streets of historic Salem.

Since shipbuilding and whaling were two of New England's most important commercial activities, it is not surprising that a great deal in general has been written about these maritime pursuits. It is a paradox, however, that with a large number of Salem people engaging in the whaling industry, so little of it is apparent in the abundant literature on the Salem theme, which for the most part was dominated by the fame of Nathaniel Hawthorne, whose writings were not particularly nautical, albeit his father, Captain Daniel Hathorne, was a seafaring man and he himself was with the Salem Custom House at the head of Derby Wharf. Nevertheless, while it is true that years after a whaling voyage, some whalemen had the urge to commit to paper something of their incredible experiences, most writings on Salem whaling were confined to bare statistics compiled for susceptive investors and published in Salem newspapers along with the usual listings of whaling vessels, tonnage, barrels of oil secured, arrivals and departures.

Yet, in the nature of things, it was inevitable that at some time a story of great power would come to light about Salem's hidden whaling era. Eventually the proper recognition of the literary value of a cache of unpublished whaling logs and journals was to reveal the fascinating whaling adventures of this romantic seaport. It so happened that it was my own particularly good fortune to be privileged—through access to the fragile fading and almost illegible sea journals—to get at the heart of Salem's whaling story. This I have related against the background of New England and world whaling.

It is interesting to note that a knowledge of the long forgotten whaling voyages, the officers and crews, their hardships, dreams, triumphs and failures sheds new light on the Pacific area as we know it today. When World War II trained its strong searchlight upon the Pacific Ocean, focusing upon the familiar old whaling grounds of the eighteenth and nineteenth centuries, thousands of Americans became newly aware of island names once known only to the whaling clan, to sailors and to traders of an early America. With Alaska and Hawaii now States in the Union, with missile testings, naval bases, conflict of national interests concentrating in what was once the whaleman's private paradise, his secrets come tumbling out of Davey Jones' well-guarded locker.

Although Salem, Massachusetts retains its position as a focal point of American history, it is my feeling that the decision of my publishers to re-issue *Whaling and Old Salem* was not primarily for this reason nor for other relative points of current interest. I believe they wished to illuminate the period of whaling by means of the wooden sailing vessels in juxtaposition to today's modern whaling industry, as the whale hunt appears on the industrial rosters of several countries in a format barely recognizable when contrasted to the age-old whaling rituals.

For some time now the lumbering old whalers—upon whose decks was enacted much of the drama of whaling—have had their day, and some believe that all whaling has therefore vanished from the scene. As many are aware, this is far from being the case inasmuch as pelagic whaling continues unabated in full strength season after season in the Antarctic. An opportunity, therefore, to compare the modern and the old was most inviting.

In the half century which has elapsed since Roald Amundsen of Norway first reached the South Pole, political, scientific and transportation events have brought the frozen southern continent into the arena of world affairs. Long recognized as the greatest of scientific laboratories, Antarctica has now by historical coincidence become the only continent with treaty guarantees of peaceful purposes and friendly international cooperation, with international attorneys observing any precedents which may be applicable to the evolution of international space law of our atomic age. Little time is being lost by interested nations. At the close of 1961, expeditions from ten countries were occupied in Antarctica with explorations, mapping and scientific projects. In June of 1961, the Antarctic Treaty was ratified by Argentina, Australia, Belgium, Chile, France, Japan, New Zealand, Norway, the Union of South Africa, the Soviet Union, the United Kingdom and the United States. Territorial status quo was set, military activities forbidden, peaceful cooperation pledged for scientific cooperation over a period of at least thirty years. By these means, a threatened race for territorial control in Antarctica was averted.

Tight in the grip of an ice age, yet of enormous strategic value (since it is in the middle of the shortest air course between Australia and South America), Antarctica, which has defied explorers by its incredible dangers, is at last yielding up its secrets to the powers of modern technology. In the southern hemisphere, the South Polar region of Antarctica lying below 60° latitude, is divided into sectors which are situated below the African, Australian and South American continents. Six million square miles in size, it is the equal of Mexico and the United States combined, and yet less than one hundred square miles is free from a permanent covering of ice. In this unique isolation where mountains rise a mile heavenward, there are no human inhabitants, no trees, few plants so that the many species of delightful penguins share the realm with the lordly whales.

In Antarctica—confined to the American and Australian sectors—whaling is currently the sole industry, linked by history to the early whaling and sealing adventurers out of the New England seaports, who were the first to bring back with them knowledge of Antarctica. The territorial claims of the United States are in fact based squarely on the whaling voyages of Captain Nathaniel B. Palmer (born in Stonington, Connecticut in 1799). In his Sloop *Hero* (44 tons), Captain Palmer found the first of Antarctica to be subsequently mapped and named Palmer Land.

As New Bedford was the queen of whaling ports of yesteryear, Sandefjord reigns today as the whaling center of the world with half of the entire industry being Norwegian and two-thirds of Norway's whaling ships and whaling men coming from Sandefjord and its vicinity. Sandefjord lies along the windings of Norway's famous Oslo Fjord, the route by which the ancient Vikings returned to their homes a thousand years ago with nearby Tönsberg the capital of Viking Norway. So the story told in *Whaling and Old Salem* comes full circle: from Ochter of Helgoland, who took whales beyond the North Cape in the White Sea in 890 A.D. to Svend Foyn, Lars ("Lars the Devil") Anderson, and the new crop of gunners and whale hunters who gain whaling fame and fortune in the frozen southland.

I am grateful to all of the whaling devotees who assisted me, as well as the authors of the works consulted. Acknowledgment is made to the Old Dartmouth Historical Society and the Whaling Museum at New Bedford, the Nantucket Whaling Museum, the Marine Museum at the Old State House in Boston, the Francis Russell Hart Whaling Museum at the Massachusetts Institute of Technology, the Provincetown Historical Museum on Cape Cod, the Whaling Museum at Sag Harbor, New York, the Seaman's Institute of New York, the Peabody Museum and the staff of the Essex Institute in Salem, Massachusetts, especially Mrs. Amelia E. MacSwiggan; the Marine Museum at Newport News, Virginia, the Marine Museum at Mystic, Connecticut, the United States Department of the Interior and the International Bureau of Whaling, Sandefjord, Norway as well as the Boston Library and the New York Public Library.

Appreciation is warmly expressed to Mr. James W. Harbin, Jr., to Mr. Henry Beetle Hough, to Mr. H. Harrison Huster, Mr. R. G. Davis. Particular thanks are due to my husband Peter who over the years faithfully accompanied me to the scattered marine museums along the New England coast, the old whaling haunts on Cape Cod, Nantucket, Martha's Vineyard, Block Island, New London, New Bedford; Hull and London in England. Appreciation is due my aunt Magdalane Baer, who supplied the subtitle and was of great personal help to me while the whaling story was in progress. Very special gratitude is due the staff of Fountainhead Publishers and particularly to Rose P. Portugal for her patience and coordinating skill in the preparation of this book.

Frances Diane Robotti

October 15, 1962.
New York City

Publisher's Preface

With the trend of publishing books which point up the historic past of the United States, it is with pride that we re-issue *Whaling and Old Salem (a chronicle of the sea)*, a definitive study of America's most picturesque of industries out of Salem, Massachusetts.

Salem, renowned for its trade with exotic world markets, was added to the roster of New England whaling ports when Mrs. Robotti's book first appeared a decade ago. Rounding out the American whaling record, it was widely acclaimed in popular and scholastic circles. When the University of Illinois selected *Whaling and Old Salem* for inclusion in the Encyclopedia *Our Wonderful World,* it became an American whaling classic.

Perhaps the most dramatic aspect of this book is the author's use of the original whaling logs and old sea journals which she brought to light from the archives of the Essex Institute in Salem. Here, in a notable contribution to whaling literature, is a compilation of first-hand, on-the-scene accounts as told by participating captains and mates of the whaling game as lived by officers and crews from day to day, recaptured from the old sea note books of more than a century ago. Adventurous spirits will be rewarded by the rich lore of the sea within these pages.

As pointed out by Henry Beetle Hough, the distinguished New England author and whaling authority, Mrs. Robotti has with characteristic astuteness gone to the heart of the great maritime industry in its many phases up to and including the modern pelagic whaling prosecuted today in the Antarctic; covering as well the geographical spread of the whalers from the Atlantic to the Pacific Coast, pointing up particularly the effect of the California Gold Rush on the whaling picture at San Francisco, on to the vast whaling theater of the South Pacific Ocean, and enumerating the many nations participating in the whale hunt.

The author has completely revised and enlarged the text with knowledgeable skill. In its new presentation *Whaling and Old Salem* is tantamount to a personal sojourn among the mysteries of a whaling voyage, vividly recreated to entertain and to familiarize the reader with a robust early America in its adventures to far places of the world.

Introduction

By HENRY BEETLE HOUGH

In the year 1835 the second most populous place in Massachusetts, ranking after Boston, was Salem, and the third was Nantucket. It happened, too, in the words of Charles Francis Adams, the historian, that this same year saw "the birth of the world as we know it"—trains ran for the first time from Boston to Lowell, Providence, and Worcester. Industrialization and urbanization had begun their irresistible march.

Until then the central orientation had been toward the sea, and the triumvirate—Boston, Salem and Nantucket—represented the main theme, tradition, and prosperity of Massachusetts and, with other ports, of all New England. As a new era begins, an old one must end. Boston turned to manufacture, and Salem and Nantucket were never again to be so great.

In retrospect, each of the famous old ports has a distinctive image. Salem is associated with square-rigger and clipper ships, the China trade and the opening up of markets with St. Petersburg, Zanzibar, Sumatra, Calcutta, Bombay, Batavia, Arabia, Madagascar and Australia; Nantucket is associated with greasy voyages of the whaleships. Though they are accurate in a symbolic sense, these images need considerable amplification. Nantucket, for instance, though it went early after whales, also was early in dropping the industry; no Nantucket vessel ever went into the Arctic.

And Salem, as Mrs. Robotti shows in this book, in addition

to its other maritime interests, was by no means an inconsiderable factor in American whaling. Salem could even claim a priority over Nantucket in point of time, though most Salem whaleships were to sail in the early and middle years of the nineteenth century when American whaling attained its greatest prosperity.

If the year 1835 could be taken as a year of decision, a sort of dividing line, still the sea not only continued to hold its importance but even to gain importance through the pursuit of whales in the period that followed; for, ironically, it was the whaleships that provided the only means of illumination and of lubrication for the new cities and the new industries, until the final exploitation of petroleum.

In presenting Salem's whaling history in its full dimensions, together with annals that help to make the era come alive again, this volume is an important contribution to whaling literature.

Mrs. Robotti not only establishes Salem firmly in its proper place in respect to whaling, but provides a work that is informative as to the whole history of the pursuit and as to the ways of whaling and whalemen. Anyone who is interested in maritime history or in the flavor of seafaring times will find these pages rewarding.

Edgartown, Martha's Vineyard,
Massachusetts.

Table of Contents

Dedication			v
Author's Preface and Acknowledgments			vii
Publisher's Preface			xiii
Introduction			xv
List of Illustrations			xix
CHAPTER	1	Early History of the "Royal Fish"	1
	2	Did Salem Whale Before Nantucket?	13
	3	Off-Shore to Deep-Sea Whaling	18
	4	Salem and the Fur Seal Fishery	26
	5	Whale-oil, Whalebone, Whale Beef	39
	6	Salem Begins in Earnest	55
	7	Whaling Merchants, Officers and Crews	61
	8	Brig, Schooner, Bark and Ship	71
	9	Merchant Service or Whaling Cruise?	88
	10	George Dodge of Salem Ships on a Whaler	95
	11	The Try-Works: From Blubber to Oil	110
	12	Lengthy Voyages, Scrimshaw and Gams	115
	13	A Whaleman's Lay from an Agent's Daybook	127
	14	Romance in the South Seas	138
	15	On the *Emerald* to the South Pacific	152
	16	On the *Reaper* and *Izette*	167
	17	Salem's Greatest Whaling Year	180
	18	Shipwrecks, Disasters, Explorations	184
	19	The California Gold Rush	194

20 Whaling Goes West to San Francisco 200
21 The Civil War and Whaling's End 206
22 Salem's *Blubber Hollow* 218
23 Modern Antarctic Whaling 221
24 Can the Whale Survive? 240

Bibliography 249

Appendix
Massachusetts whaling statistics 259
Some comments on whaling pictures 263
Herman Melville and *Moby Dick* 267

Indices 275

List of Illustrations

Following page 104

Whaling off-shore on Long Island, New York. Drawn by W. P. Bodfish. Author's collection.

"The Capture" From the Allan Forbes Collection. Lithograph of Sperm Whaling after drawings by A. Van Best and R. S. Gifford, corrected by Benjamin Russell. Endicott & Co., New York, 1862.

"In a Flurry" From the Allan Forbes Collection. Lithograph by N. Currier, New York, 1852.

Old whaling agreement dated 1743/4, Boston, Massachusetts. Collection of H. Harrison Huster, N.Y. Member of Nautical Research Guild.

Vessel rigs. Drawn by Charles G. Davis, naval architect and famous ship modeler of the Frigate *Constitution,* the *Sea Witch,* the Whaler *Sunbeam* and others.

Ship *Emerald* of Salem, Water color by M. Macpherson of original by G. M. White.

Ship *Bengal* of Salem in company with the *Foxhound* of England, Pacific Ocean whaling grounds.

A whaling station on the California coast by Frenzeny. From Harper's Weekly (Journal of Civilization), New York, June 23, 1877.

Scrimshaw on whale's tooth, front and reverse. *Homeward Bound.*

Scrimshaw of whaler flying the British Jack in ancient Capernum, Sea of Galilee. H. Harrison Huster Collection.

Charles W. Morgan, last American whaleship. Marine Museum, Mystic, Connecticut. Photograph by Herbert H. Corey.

Model of Whaling Brig *Viola.* H. Harrison Huster Collection.

South Sea whalers boiling blubber by O. W. Brierly. Illustrated London News, July 1, 1876. J. Geenaway.

The Eastern Hemisphere as depicted in world map dated 1840. Published by Henry Teesdale & Co. London, England.

Sailing from New Bedford, Nov. 15, 1861, of the *Stone Fleet.* Lithograph by L. Prang & Co., 1862, in the Allan Forbes Collection.

Clifford W. Ashley (*The Yankee Whaler*) mentions some 60 different items of Scrimshaw in four categories: ornamental, utilitarian, personal trinkets and play.

Whalebone swift, used in winding yarn for knitting and spinning. H. Harrison Huster Scrimshaw Collection.

Whalebone busks, frontal center stay of a woman's corset. Jagging or crimping wheels for marking the edges of pies or tarts.

Shipwreck of the Whale Ship *Essex* of Nantucket by a sperm whale depicted in scrimshaw on a whale's tooth. H. Harrison Huster Scrimshaw Collection.

Model of a Fiji Island temple. Kindness of the Peabody Museum.

Cannibal bowl from the Solomon Islands. Kindness of the Peabody Museum.

Moorea, Tahiti—Peak over Cook's Bay.

Tahitian Vahine doing her daily laundry.

Couple dancing in costumes of the Island of Bora Bora, Tahiti. Kindness of Air France: *Tahitian Holiday.*

Rajendra Dutte, a merchant of Calcutta. Life-size portrait figure. In Peabody Museum, Salem, Massachusetts.

Life-size figure of Canton merchant. Head and hands were carved by Salem's Samuel McIntire. In Peabody Museum of Salem.

Gilbert Island warrior in coconut fibre armor with rare gauntlets set with shark's teeth. Life-size figure in Peabody Museum of Salem.

Hawaiian idol, six feet tall, one of many found by white men in a sacred island enclosure. Only three are in existence: British Museum, London; Bishop Museum, Honolulu; Peabody Museum, Salem.

Following page 168

The whale compared to other animals from an old drawing by W. Croome engraved by J. Andrews. Very rare old print. Author's Collection.

Arctic whaling scene after Currier & Ives from oil painting by Natalean and Noble Cassady of California.

Captain William Ramsdell (1766-1842) of Salem, Master of the Salem Ship *Lydia.* Taken from the album of the Salem Marine Society. Kindness of the late Colonel Arthur Treadwell Dalton, a Salem member.

Metal and glass whale-oil lamps. Essex Institute, Salem.

Arctic perils. The fleet enclosed in the ice-pack. The camp on a grounded iceberg.

Escape of the Bark *Florence.* "Recent disasters of the whaling fleet" by Captain W. H. Kelley. Harper's Weekly, November 18, 1876.

Eskimos whaling off-shore and fishing by spearing under the frozen surface. By Roberto Freschi in an ancient Italian volume. Author's collection.

The Bark *Wanderer,* wrecked on the Island of Cuttyhunk August 25, 1924, out of New Bedford, by the famous marine artist Gordon Grant. Collection of The Seamen's Bank for Savings, New York City.

The British Whaling factory-ship *Balaena* active in the Antarctic. Drawn by G. H. Davis, long famous for his accurate marine art studies, Chief Staff Artist of the Illustrated London News.

Designs by Svend Foyn of Norway. Kindness of James W. Harbin, Jr., Nautical Research Guild.

Salem Custom House in 1850 at the head of Derby Wharf.

Benjamin Webb—Salem whaling agent.

Nathaniel Hawthorne's stencil for marking merchandise at the Salem Custom House where he conceived his great novel *The Scarlet Letter.*

Whale-oil lantern.

Holder for spermaceti candles.

Front doorway of the Pingree House on Essex Street and elegant hallway.

The old Candle Factory on Nantucket Island.

Whaling Ship try-works, Nantucket.

Doctor's boxes contained every conceivable remedy, a welcome purchase for a long voyage.

Merchant Princes of Salem's great age of sail, whose investments included whaling voyages to many world whaling grounds: Elias Hasket Derby, William Orne, William Gray, Joseph Peabody.

Bark *Reaper* of Salem. Built in Amesbury in 1820.

Ship *Eliza* of Salem, altered to a Bark in 1838.

Dining room of the beautiful Pingree House and rear parlor. The quiet elegance of the interior and furnishings depict the gracious era when Salem held a prominent place in the maritime industries of New England.

Master engraving from the original design of Adriaan Van de Venne, Jacob Cats, Amsterdam, Holland, J. Rotelband (1712). Appears in Volume 1 of 2 Volumes. The Dutch learned the art of deep-sea whaling from the early Basques. Courtesy New York Public Library.

WHALING AND OLD SALEM

CHAPTER

1

Early History of the "Royal Fish"

Oh, the rare old whale mid storm and gale
In his ocean home will be
A giant in might, where might is right
and King of the boundless sea!
—Whaling Song.

From ancient times whaling challenged the boldest and hardiest of seamen in a calling full of danger and profitable only by the whims of chance. The origin of the whale hunt is lost in antiquity. Men hunted whales before recorded history. In Europe's Stone Age aborigines took whales, extracted their oil and used their meat. Genesis tells of God's creation of "great whales." In addition to the familiar tale of Jonah and the whale, there are biblical references in such books as Isaiah and Psalms. From the Hebrews and Greeks to Fiji Islanders, nations have had a name for the Titan of the deep which reflects the ocean's own power. Nor has literature overlooked the whale. There are many allusions to the world's largest creature.

The Basques along the north coast of Spain are generally spoken of as Europe's first whalemen. It is true that they were the first people to specialize in the whale's capture. Not to be forgotten are the efforts of the Vikings. An epic feat of whale hunting was performed by a northman named Ochter, a native of Helgoland. Ochter's achievement won for him the distinction of being the first recorded whaleman. The story goes that this Norwegian voyaged north, skirting Norway as he sailed beyond the North Cape to the White Sea. His route was that which the whale hunters "commonly used to travel."

Ochter afterwards described his journey to King Alfred of England. This monarch of ancient times was particularly interested in the icy regions and preserved a knowledge about them in his books. The amiable and literary king wrote down Ochter's whaling experiences in Saxon with his own hand from Ochter's lips. It appears that six stout-hearted men under Ochter's leadership killed sixty whales in "his own country in two days." Richard Hakluyt, the sixteenth century chronicler, places this miracle at 890 A.D. Adam Anderson's *Origin of Commerce* says it happened three years earlier. Centuries before Alfred the Great and Ochter, the Persians are known to have practiced off-shore whaling in the Arabian Sea.

The early efforts of the Norwegians, heroic as they were, are to be regarded in the same light as the fishing expeditions of other early peoples such as the Japanese, Eskimos and our own Indians of New England who killed blackfish and Right Whales around Cape Cod before the coming of the English, then gave their skilled services for 3 shillings a day, paid in oil. The first people to make of whaling a systematic, regular business were the Basques in the villages on the coast of the Bay of Biscay. The Basques of the twelfth century originated methods still in use today in greatly modified form. Their word *arpoi* meaning to take quickly, or to grab, became *harpoon*. The Basques built watch towers along the coast to spot the whales, then set out in small boats to capture the mammal.

In his *Story of the New England Whalers,* John Spears points to ancient documents which indicate that in 1150 A.D. King Sancho the Wise granted special privileges in the matter of whaling to the City of Sebastian. In 1261 A.D. a tax was levied on the importation into Bayonne, a port of Biscay, of the whale's tongue which was considered a great delicacy. By the thirteenth century, whaling flourished for the Basques who hunted the Black Right Whale, a species now practically extinct. Constant pursuit decreased their numbers until the Basques had to go as far as Newfoundland and Iceland to find good hunting. They are said to have reached the banks of Newfoundland in 1372, where they found a new species of whale later called the Greenland Right Whale.

The Basques taught both the British and the Dutch the art of deep-sea whaling. In the fourteenth century the Britons were already active in the northern whaling grounds where they were soon followed by other nations. The Basques reached the Gulf of St. Lawrence and pressed on to the edge of ice off Greenland where they led the way for the Icelanders and Spaniards. After some five hundred years of whaling, the efforts of the Basques declined in the seventeenth century, some writers say in the sixteenth. Crowding north, whalemen from Scotland and England headed a thriving industry by the middle of the sixteenth century, with permanent shore stations for extracting oil.

The chief rivals of the British were the Dutch. A native of the Netherlands is credited with the discovery of Spitzbergen with its numerous whale herds. In searching for a route to the riches of India, William Barentz in 1596 traveled in the high north latitudes and found the group of islands in the Arctic Ocean between Greenland and Nova Zembla known as Spitzbergen. The coasts were covered with immense glaciers and chains of mountains rising to a height of four thousand feet. The bays and fjords were alive with whales. Barentz died in Nova Zembla in 1597, but some of the surviving crew reached home with the news that they had found numerous whales.

The report spread through oil-hungry Europe. Soon English, French, Dutch and Spanish ships were actively engaged in the hunt, each jealous of the other, each eager to elbow out a competitor. The tension between the English and Dutch broke into open hostilities, the English basing their claim to Spitzbergen on the prior discovery of the Englishman Willoughby in 1553. Whaling proceeded in spite of the war. As the Black Right Whale of the Basques disappeared after so many years of ruthless harpooning, the pursuit of the new Greenland Right Whale began in earnest. This is placed at the year 1611.

The Dutch kept many whaleships in the Arctic as their profits grew. Their headquarters for *trying-out* their catch were a few degrees from the North Pole. This permanent settlement replete with tailors and shoemakers was called Smeerenburg or Blubbertown. A thousand ships visited this place during the year. Whaling was done during the four summer months in which the sun shines day and night. All the valuable space in Dutch vessels was used by evolving the practice of extracting the whale's oil and whalebone on location instead of transporting the blubber to the home port. For this purpose boilers and tanks were installed on shore with coolers and all the necessary equipment of a factory. By 1612 the Dutch had acquired so many experienced whalemen and so much skill that they were in the lead of all their competitors. During their long years of herring fishing, the Dutch had built strong seamen and a splendid fleet of ships which they now turned to the profitable whaling industry. Other nations also rushed to the new whaling grounds and quarreled and fought for the spoils. This unpleasant situation existed until 1618 when the Dutch Whaling Company sent out seven armed ships to Greenland which resulted in the division of the northern area among all participating nations to their mutual profit.

In 1630 England sent out many whaling ships and her early whaling history is filled with tales of shipwreck, isolation of whalemen on ice floes, and starvation. Still the rewards were great enough to make such risks worthwhile. It was the Dutch, however, who established an undisputed leadership in this field to which the whale owes its name. The German or Dutch word to roll or to wallow was *wallen*. Whale-oil is still called train oil from the German word *tran* meaning fish oil. The removal from the carcass of the whale's blubber, a solid layer of fat under the skin, is called *flensing,* the Dutch word for this process. By 1680, the Dutch provided the chief oil supply of

Europe with an active fleet of 260 vessels and 14,000 seamen employed in the work. In 1697, the number of Dutch vessels was reduced to 188, but they carried home the extracts of nearly 2,000 whales.

From 1598 the port of Hull, England, had regularly sent whalers to Iceland and the North Cape, and to Spitzbergen in the seventeenth century. Even if he were a warm-blooded mammal, who had adapted himself to living in the sea, the English called him a *fish,* and the enterprise of his pursuit the *whale fishery*. By early law, all whales either thrown ashore or caught near the coast became the property of the king. In consideration of the sovereign's guarding and protecting the seas from pirates and robbers, he was to receive as a tenth branch of his ordinary revenue the right to *royal fish,* which, says Blackstone, "are whale and sturgeon."

When the whale was first discovered in England, the King became the Honorary Harpooner and by law received the head and the Queen the tail of all whales captured along the English coast. As there is very little left of a whale between the great portion occupied by the head and that by the tail, this division was like halving an apple. It left practically nothing for the whale fishermen. Later it became traditional for the government to receive one-third of every whale caught instead of two-thirds. In 1662 the Church of England also put in its bid for a part of every whale. All these demands for portions proved discouraging to whaling efforts. By 1702 Queen Anne found it necessary to encourage whaling by offering a bounty for catching whales.

Finally recognized as a valuable national enterprise, English whaling was subsidized by a bounty system at the beginning of the eighteenth century and faithfully continued until 1824 when bounties ceased. Great amounts were expended over this long period. For instance, between 1750 and 1769 the government of England spent £613,261, a large sum considering the value of money at that period. This kind of encouragement allowed the English to overtake and finally to surpass the Dutch. By 1812 the once-mighty whaling fleet of the Dutch was reduced to only sixty sail.

Captain James Cook first reported whales in the southern seas in 1775 shortly before his discovery of Hawaii then known as the Sandwich Islands. After this, American and British whalers began their profitable voyages thither. The British turned whaling into a paying venture toward the end of the eighteenth century when they acquired a more basic knowledge of whaling from the Americans. In 1786 some of the whalemen of Nantucket Island were induced to emigrate to Nova Scotia to teach the inhabitants the mysteries of whaling. The whaling colony they helped to found was called Dartmouth. Also by request, Nantucketers settled in Milford Haven on the coast of England to instruct the English in the art of whaling.

Whales were pursued from one habitat to another and in the process explorations and discoveries took place. As for whales in the New World, Captain John Smith of Virginia fame found them so plentiful along the coast of New England that he turned from the original object of his voyage in order to give chase. Colonial planners included whales in their considerations of advantages in settling the American wilderness. Whales are mentioned among "other merchantable means to raise profit to the industrious inhabitant or diligent trader" in the *Brief Relation of the Discovery and Plantation of New England* published in London, England, in 1622. In the same year Mourt's *Relation* was printed in London and advised prospective colonists of the New World that "Whales of the best kind for oil and bone are said to abound near Cape Cod which for that reason is spoken of as likely to be a place of good fishing."

These reports were released one year before Roger Conant was placed in charge of the Cape Ann settlement in 1623 by the Dorchester Merchants of England, and four years before Conant's group came down from Cape Ann to begin the settlement which became Salem, Massachusetts. When the *Mayflower* of the Pilgrims arrived in 1620, and the *Arbella* of the Puritans ten years later, the colonists saw whales sporting about their ships. It was not surprising, therefore, that the royal charter of the Massachusetts Bay Colony secured from Charles I in 1629 should include a guarantee of the right to catch whales within the waters of the proprietors. Not unprophetically was a vessel in Governor John Winthrop's fleet of 1630 called *The Whale.* It is entirely possible that this ship was once engaged in whaling from which it may have taken its name.

Before sailing on her eventful voyage to Plymouth, the historic *Mayflower* had apparently been a whaler. After transporting the Pilgrims to America, the *Mayflower* was engaged for a number of years in the thriving Greenland whale fishery. In 1620 when the colonists found an abundance of whales "playing hard by the *Mayflower,* many of them were eager to undertake their pursuit especially as there were among the ship's company a number of men skilled in the Arctic whaling, a much more difficult profession" than the pursuit of whales in the calm waters of Plymouth Bay.

Three hundred years ago the whole New England coast was haunted by the peaceable and rather slow-moving Right Whales which were present in especially large numbers during the winter months. They were often seen in Buzzard's Bay and all the length of Long Island Sound. When Richard Mather came to the Bay Colony in 1635, he saw what he described as "mighty whales spewing up water in the air like the smoke of a chimney of such incredible bigness that we did not wonder that the body of Jonah could be in the belly of a whale." Mather made the common mistake of many who viewing the whale from afar seem to see a fountain of water rising. The whale's great breath condenses when it strikes the atmosphere above the surface of the sea and forms a cloud of steam. You can tell from his spout what kind of a whale he is. Mather's whale was a Right Whale whose breath comes in twin jets shooting almost twenty feet in the air, then becoming progressively shorter. The columns rise straight upward, divide on top and fall over like the drooping branches of the willow tree. The breath of the Sperm Whale is a forward-slanting spout from a single nostril which presents a thick, curled bush of white mist continually rising and falling away to the leeward.

Mather's second observation bearing on the biblical legend of Jonah has long been a moot question. Strange as it may seem, the throat of the world's largest animal is no bigger than a man's arm. The only exception is the toothed Sperm Whale which feeds on giant squids or octopuses and has a throat large enough to take care of his choice of food. The Sperm Whale which could swallow a man, and has swallowed ten-foot sharks, is—interestingly enough—reported to be from the Mediterranean. It is probable, however, holds the naturalist Ivan T. Sanderson of New York "that the biblical story about Jonah derives from some much more ancient tale connected with early whaling in the Arabian Sea or Persian Gulf."

To return to early whaling in the Massachusetts Bay Colony, whale-oil was a staple commodity there as early as 1642 according to Joseph Felt, *History of Ipswich.* In 1644 Long Islanders in New York had regulations about the drift whales cast up on their beaches. In any discussion of the beginnings of American whaling, mention should be made that the first organized prosecution of the whale fishery by Americans was by the settlers on the eastern end of Long Island. American whaling originated quite humbly like that of the ancient Basques as a coastal and in-shore fishery. By the middle of the seventeenth century, offshore whaling was in progress from Long Island to the St. Lawrence with Southampton playing an important role. As soon as the Long Islanders could build strong enough whaleboats, they rowed out after the whales and captured them a little offshore. The Salem chronicler Joseph Felt reminds us that whaling became the first employment which engaged the attention of the colonists at New Amsterdam and at Delaware.

Attempts to regulate whaling matters early found their way on the statute books of New England towns. In early years dead whales were so often cast upon the shores that laws and regulations for deciding their ownership were constantly being enacted. In the town records of Orleans, Cape Cod, we find that the division of whales or part of a whale was provided for. "At a meeting this 8th day of December 1662, it is ordered that every whale that shall come within the bounds of this township after the date aforesaid, shall be disposed of viz: the one half of this town to have one whale and the other half to have the second, and so for the future, so also it is already lotted that half of the town that lies on the north side shall have the first whale."

Called the "Royal Fish" of English monarchs, the whale is nevertheless a warm-blooded mammal. The ocean is his home, but he is *not* a fish. The Germans called his oil *tran,* or fish oil, and the subject has often arisen to torment those who traffic in whale products. Over a hundred years ago, it was decided in the State of New York that from a legal standpoint the whale is a fish. In those years, a tax on fish-oils was levied by the state. The importers of whale-oil refused to pay the tax on the ground that a whale was not a fish. The decision on a trial case held that for purposes of law a whale *is* a fish, and its oil was therefore fish oil and subject to the tax. This type of hairsplitting was recently found in Victoria, British Columbia, where certain butchers refused to handle whale steak on the ground that it was not meat. They scornfully maintained that the proper place to sell this product was a fish market. This objection is the exception rather than the rule.

There is little general knowledge in this country about whales and whaling and many popular errors exist. In 1918 a widely read magazine said a whale had been killed that "naturalists believed to be 500 years old." In 1934 the error was repeated, or doubled, and the whale's lifetime referred to as 1000 years. Likewise in 1934 the amount of ambergris a whale could produce, as well as its source, was still generally in doubt. Until comparatively recent years, little was scientifically known about the breeding and migratory habits of the world's largest mammal. There are, to be sure, many whalemen who are wise in the ways of their quarry, but few are Melvilles and this rich whaling lore may never reach paper.

British investigations began in the Antarctic in 1925. In 1934 British scientists admitted it would be twenty years or more before the life story of the sea Titan could be written. In 1947 the United States with Antarctic possessions of its own was reported to be following the lead of Great Britain in making scientific studies on which long-term policies could be based. Much research has been necessary on the meteorology of the Antarctic region, the most fruitful of all whaling grounds, and on the habits and migratory movements of the whales to make the work of the whaling expeditions in these far southern seas safer and more productive.

Before the war, the work was chiefly undertaken by the Discovery Committee of Britain which focused its attention on the natural history of whales. Three ships carried on the work, the Royal Research Ship *Discovery, Discovery II,* and the *William Scoresby.* A marine station was established at South Georgia near one of the shore-based whaling stations and detached parties worked in the factory ships and other stations where the anatomies of many thousands of whales were examined in the difficult study of breeding and growth. When war broke out, work at sea was suspended and little could be done on the accumulated material. In 1945 scientific research and survey work was resumed on the Falkland Islands, some of the most remote of British Dependencies with N. A. Mackintosh, Director of Research in Britain to the Discovery Committee, which issued Discovery Reports.

In general, it was learned that whalebone whales breed slowly, but mature comparatively quickly. Of the three important species, the medium-sized Finback Whales, average length 66 feet, were still plentiful in the Antarctic Ocean, but the stocks of the great Blue Whales averaging 75 feet, and the smaller Humpback whales, averaging 54 feet, were reduced as a result of the intensive hunting before the war from which they had not recovered. The whale herds were hardly more numerous than before the war. This did not surprise the British scientists who sailed with the fleet. They had warned that whales were sluggish at multiplication.

Gestation period for the whale mother is not less than 12 months and she bears one baby every two or three years on the surface of the sea and suckles it on warm milk for a period of 7 months to a year, depending on the species. Many of the cows which are killed have been pregnant as there is no way to tell their sex or condition. The slow, unproductive rate and comparatively short life, handicap the whale in his fight for survival against the incursion of the highly efficient business whaling has become.

The length of the largest Blue Whale registered was 100 feet. At birth a baby Blue measures over twenty feet, which makes him larger than any existent full-grown land animal. For sheer size the Blue or Sulphur-bottom is larger than any prehistoric giant known to science. The giants of the Age of Reptiles, which began some 150 million years ago, did not measure up to the largest Blue. Artists who take poetic license often depict the Dinosaurs as contemporaneous with ancient man. Actually they were extinct 70 million years before man evolved. Of the Dinosaurs, the Megalosaurus measured only 20 feet. The Tyrannosaurus Rex, fiercest and most terrible of Dinosaurs, was 18 feet high and measured 47 feet from his head to the end of his tail with jaws 4 feet long. The Brontosaurus, whose great skeleton has been unearthed in the State of Wyoming, weighed 40 tons. The animal which seems to compare most closely in size to the largest of the Blue Whales was the Zeuglodon. The Zeuglodon, which some 40 million years ago swam the seas over what is now southern United States, measured 80 feet in length.

Some idea of the whale mammoth of our own times in girth and proportion of flesh, bones and blubber can be secured from considering the model of a Blue Whale in the National Museum at Washington, D.C. This Blue is 78 feet long. In life, he was 35 feet around the shoulders and had a head length of 19 feet. The flukes were 16 feet across and his total weight was 63 tons. The bones weighed 8 tons, blubber 8 tons, flesh 40 tons, whalebone 7 tons. A Blue Whale of 89 feet length weighed 119 tons and yielded 166 barrels of oil. The weight of the largest Blue Whale captured, which was 100 feet in length, has been given as 120 tons, equivalent to the weight of 2,000 people, or 20 elephants.

In absolute strength, the whale is the strongest creature on earth. The muscles of a Blue Whale weigh over 50 tons. When he travels at the rate of 10 knots an hour, the whale develops about 47 horsepower. He can tow a killer boat up to seven hours before reaching exhaustion. In their deep dives and leaps, the whales perform impressive feats of strength. When *breaching,* the Finner Whale can hurl his 70-ton body many feet clear of the water. It is customary for the Sperm or Cacholot Whale to feed at depths of 1500 to 1600 feet. A 45-foot Sperm was found entangled in a broken cable off British Columbia at a depth of 3,240 feet. At this depth the pressure on its back equalled the combined weight of the ocean liners *Queen Mary* and *Queen Elizabeth.* Whales have been known to wreck whaleships and probably some ships which have disappeared at sea were wrecked by this leviathan. In 1820, the American whaleship *Essex,* Captain George Pollard, was rammed twice by a Sperm Whale and sank in ten minutes.

Here then is the mystery and the glory of whaling: man against the largest and strongest creature that ever lived, in which he proves that man is the victor.

CHAPTER

2

Did Salem Whale Before Nantucket?

"A bone for you to pick on"
James Calfbill (1565)

By 1664 the settlers at Cape Cod reported large profits from whale catching, and Salem, too, took its share of whales, the infant industry constituting a goodly part of the town's economy. It is customary for histories of Massachusetts and New England to omit any reference to Salem whaling in spite of its long record of active participation. There is evidence of a whaling industry of some strength being prosecuted in Salem in the fifty-year period from 1820 to 1870. The early years from its founding in 1626 until 1820 were not idle ones in a town which looked to the sea for a livelihood and profit. By 1670 Boston had a fleet of little pinnaces which cruised along the shore to watch for whales, nor was Salem backward in whale catching.

An amusing compensation for historical neglect comes to light in a controversy of a century or so ago which tried to establish Salem as engaged in the whale fishery *before* Nantucket, famous for its early whaling. The claim of priority for Salem was made by the Reverend Charles W. Upham, Minister of Salem's First Church. Mr. Upham is more renowned for his writings on Salem witchcraft than as a whaling authority, but his arguments in the case of whaling are learned and sound. He presented his facts to an audience in Salem's Lyceum Hall and the lecture was reported by the *Salem Register* in its issue of April 13, 1837. The *Nantucket Enquirer* and the *Boston Morning Post* promptly denied the claim. Both of these newspapers cited the contract drawn by Nantucket with James Loper and John Savage in the year of 1672. Both men were to proceed to Nantucket and initiate the whaling industry there, according to Obed Macy's *History of Nantucket.*

Mr. Upham replied by way of the press, conceding that such contracts existed bearing the date of 1672, and recalling their provisions. By the terms of the agreement, James Loper of East Hampton, Long Island, was to remove with his family to Nantucket Island and there carry on whale catching for a consideration of land and payment. John Savage was likewise to receive land—ten acres—on which to live if he and his family would establish themselves there. As a cooper, John Savage would provide barrels in which the whale-oil could be stored. As an added inducement, both Loper and Savage were to receive one-third of the proceeds and two-thirds were to go to those who embarked with them on the enterprise. Starbuck says Savage *did* remove to the Island, but Loper did *not;* that the history from 1672 to 1690 is *obscure.*

It was certain, in Mr. Upham's opinion, that this plan was *not* carried into operation. "It is incredible," he held, "that a people who have ever since shown such an unrivaled genius for whale catching could have been engaged in it for eighteen years, that is from 1672 to 1690, and still remain so unskillful as to have to send to Cape Cod for a man to instruct them in the business in 1690."

The reference to sending to Cape Cod for a man to teach them is the familiar one from Macy's *History of Nantucket* about Ichabod Padduck. "Finding that the people of Cape Cod made greater proficiency in the art of whale catching than themselves, the inhabitants in 1690 sent hither and employed a man by the name of Ichabod Padduck to instruct them in the best manner of killing whales and extracting their oil." There is no evidence that Nantucket whaled *before* procuring Ichabod Padduck. There *is* evidence that they whaled after procuring his services. "Whale fishing began at Nantucket in the year 1690" wrote Zacceus Macy in 1792.

To substantiate his claim that Salem whaled before 1690, the accepted date for Nantucket's initiation of the fishery, Mr. Upham produced manuscript letters of the Higginson family of Salem from the depositories of the Massachusetts Historical Society. The correspondence he used for this purpose has reference to the trade and business conditions in the Massachusetts Bay Colony, especially in Salem, during the latter part of the seventeenth century. It appears that Nathaniel Higginson, the son of the Reverend John Higginson and grandson of the Reverend Francis Higginson, Salem's first minister, had migrated to the East Indies about the year 1690, and there became the Governor of Fort George. He was successful in various business enterprises, amassed a considerable fortune, and intended to settle in England. His father, the Reverend John Higginson, and his brother, Major John Higginson, wrote young Nathaniel many letters for the purpose of inducing him to change his mind and establish himself permanently in Salem. They gave him reports of business and industrial matters to support their opinion that he would do very well as a Salem citizen.

Mr. Upham secured the information he required from the letters which passed among these men. In answer to the inquiries made by the brother in the East Indies of the brother and father in Salem about the various branches of trade and commerce, there was given a full account of exports, imports and of the entire business in Salem and Boston at that time. The brother in Salem, Major John Higginson, mentions as the first article of produce "Whale and other fish oil and whalebone." In another letter he writes: "For my own part in the year *1689* when this war first broke out, I obtained a competent estate being as much concerned in the fishing trade as most of my neighbors, but since that time have met with considerable losses. Trade is much decayed. Of 60 odd fishing ketches belonging to this town, about six are left. I believe no town in this province has suffered more by this war than Salem."

The war referred to was England's war with France beginning in 1689. Fishing in Salem was almost completely broken up one year *before* Ichabod Padduck went over from Cape Cod to teach the Nantucket Islanders how best to catch whales and to extract their oil. This indicates that it flourished for some years in Salem prior to 1689. That the *fishing trade* mentioned by John Higginson was partly made up of whaling is evidenced in a letter addressed by him and Timothy Lindall to Mr. Nathaniel Thomas, a Cape Cod lawyer, under date of March 18, 1691. From the contents of this letter we learn that Mr. Thomas is employed to obtain satisfaction for injuries suffered during the *two* preceding winters when the Cape Codders took possession of whales harpooned by Salem men. By colonial law the whales belonged to those who first harpooned them. The letter complained:

"Sir: We have been jointly concerned in several whale voyages at Cape Cod and have sustained great wrong and injury by the unjust dealing of the inhabitants of those parts, especially in two instances: ye first was when Woodbury and Company in our boats in the winter of 1690 killed a large whale in Cape Cod harbor. She sank and after rose, went to sea with a harpoon, warp, etc. of ours, which have been found in the hands of Nicholas Eldridge. The second case is this last winter, 1691. William Edds and Company in one of our boats, struck a whale, which came ashore dead, and by ye evidence of the people of Cape Cod, was the very whale they killed. The whale was taken away by Thomas Smith of Eastham and unjustly detained."

John Higginson
Timothy Lindall

Mr. Upham made the claim that Salem preceded Nantucket, not that Salem was the first to participate in the whale fishery. Whaling was so extensive in every seaport along the Atlantic

Coast that previous to Nantucket's entry in 1690, it was necessary to have statutes to regulate it. Laws on the manufacture of whale-oil and on whale catching appear as of 1671, 1662, 1656 and even down to 1652, which was seven years before Thomas Macy founded Nantucket. The *noble sport,* as Major John Higginson calls whaling in his letters, was, in truth, known from New England's beginnings.

It is not surprising that other sections of Massachusetts should have preceded Nantucket in whaling. The island was ill-adapted for offshore whaling because the shallow waters in winter were obstructed by ice. In other seasons, the whales in any numbers were prevented from approaching the shores by the shoals which extend to a great distance on all sides of the island. It was only when whaling was carried on in remote waters that Nantucket outstripped her competitors. And it was from Nantucket that the first whalers sailed to the Far East, leaving their names in many islands and ports of the Pacific, and, as Charles D. Childs observes, they also left a not inconsiderable number of descendants who did not bear their names. Many of the South Sea Islands were first discovered and named by Nantucket whalemen. After 1820 whaleships increased in size and the added tonnage sounded the death knell of plucky Nantucket's whaling industry. The island carried on in spite of the handicap until after the Civil War with the aid of *camels,* or assistants which helped to float the heavy whalers over the sand bar at the entrance of the harbor where the depth of water was only seven to nine feet.

CHAPTER

3

Off-Shore to Deep-Sea Whaling

"They that go down to the sea in ships,
That do business in great waters;
These see the works of the Lord,
And His wonders in the deep . . ."
Psalm 107:23, 24

By the opening of the eighteenth century, the products of the whale were accepted as part of the colonial economy. In a summation of foreign trade, Major John Higginson stated: "We produce here a considerable quantity of whale and fish oil, whalebone, furs, deer, etc. annually sent to England." In an almost unique recognition of Salem's whaling record, Mr. John R. Spears, in *The Story of New England Whalers,* in his discussion of minor colonial ports, makes the statement that "Salem was a whaling port of some importance as early as 1700." In considering early whaling, one must bear in mind that the whaling activities of the various American seaports in the seventeenth century and early eighteenth century were confined to whales cast up on the beaches or secured by chasing them a short distance off-shore. This type of hunting did not require vessels of heavy tonnage. It was in this class of whaling that the little ports were occupied. The records of Martha's Vineyard show that whales were killed along shore in 1702. Edgartown, the ports of Rhode Island, New Bedford on Buzzard's Bay were all active in this early off-shore whaling.

Major Higginson continued the whaling activities in which

he was engaged before 1689—evidenced in the exchange of letters with his brother in the Indies—for we find a letter of his to a friend in Ipswich written in the year of 1706 in which he discloses that he is "concerned with some others in procuring new boats for catching whales." Further proof of Salem whaling in this period was furnished by Ira J. Patch of Salem (1835-1894). Mr. Patch searched out some old testimony from the files of the Essex County Courts which referred to the "Ancient Whale Catching Customs of the Fishers in Cape Cod in 1708." The contents of these papers led to the belief that Salem and Beverly, which share the same harbor, were engaged in the whale fishery at that time.

After 1690, the year which marked Nantucket's entry into whaling pursuits, off-shore whaling was carried on widely by the New England settlers. It was the habit of the whales up to about 1748 to set in alongshore by Cape Cod. The whaling was done in the immediate vicinity of the coast and seldom beyond sight of land. There was good whaling in boats, reports Mr. Felt, and proper watchmen on the shore gave the alarm when a whale appeared. Six brave men went out in a boat with a harpoon, some lances and a line, and after the hard rowing of the chase, faced and killed the whale, and towed him back to the beach. The whole job from shoving off to return with the prize consumed but a few hours.

The all-important factor in off-shore whaling was the surf. In taking off from the beach, the surfman had to know when he could safely get off through a turbulent sea and when he could not do so. A surf of over eight or nine feet meant a whaleboat could not possibly be shoved off without meeting disaster. A man who well understands the surf and its capricious moods is Capt. Everett Joshua Edwards, a veteran whaleman of East Hampton, New York, whose ancestors from generation to generation whaled off-shore. "A whaleboat can get off in a heavy sea, if it's a ground swell and steady" he holds. "There are generally three heavy seas and then will come a slatch—two or three smaller seas behind them. You follow as near as you dare after a heavy sea, and if it's smooth behind—let her go!"

The chief difference between shore-whaling and deep-sea whaling, in Capt. Edwards' opinion, was the surf. In deep-sea whaling the whaleboat was launched from the mother ship and in a heavy sea this was a difficult enough undertaking. Outside of contending with the surf on the beach "off-shore and deep-sea whaling were basically the same. The whaleboat was the same, the tools were the same, and the methods were the same."

After being harpooned for so many years, the whales grew tired of playing a losing game in the waters around Cape Cod, and finally passed farther off upon the banks at some distance from the shore. Cape Cod and Salem whalemen followed them to their new haunts. To do this, they had to use a whaling sloop or schooner. These masted sailing vessels carried two whaleboats and a crew of thirteen men. When they overtook the whale, the two boats were lowered into the sea and the men gave chase.

Salem men secured whales in this manner before New Bedford was founded. The whale fishery was initiated in New Bedford by Joseph Russell in 1755 and within ten years, it had five small sloops. In 1775 New Bedford was well on its way to becoming an important whaling port, having 80 vessels employed in this fishery. The change from coastal to deep-sea operations necessitated whalers of heavier tonnage for the long voyages. In 1763 American whalers were to be found off the coast of Guinea; in 1765 at the Western Islands in the Azores, and two years before the American Revolution off the coast of Brazil.

Beginning the pursuit of whales so early in its history, Salem was nevertheless the last of the Massachusetts ports to send whalers to distant oceans. There seems to be some discrepancy as to these first deep-sea vessels. According to Starbuck's *History of the American Whale Fishery,* Salem's *first* whaler was the ill-starred Ship *Britannia* of 1818. Equipped for a whaling and sealing voyage, she was wrecked on Pickard's rocks off Manchester just a few hours after leaving Derby Wharf. Two earlier vessels are listed in the *Salem Ship Register* as engaged in sealing and whaling voyages to the Antarctic. Although whaling is not specifically mentioned, early voyages included the pursuit of both seals and whales. The Ship *Minerva* sailed for the Antarctic in 1799 and returned three years later. The Ship *Rachel* sailed from Salem for the same destination in 1803. If one includes in these sealing-whaling voyages Captain Frost's ship, *General Knox* of Salem, which arrived in Boston on August 17, 1784 with 600 barrels of whale-oil—the *Britannia* emerges as Salem's *fourth,* and not the *first* of Salem's deep-water whaling vessels.

Some confusion may arise from such references as that of Albert Bushnell Hart who declares in his *History of Massachusetts* that "deep-sea whaling was under way by the early 1700's." This refers not to the long voyages in distant oceans, but without doubt to the beginning made by Nantucket in the sperm fishery. In the year 1712 a fortunate accident achieved two important advances in whaling history. The story goes that Capt. Christopher Hussey of Nantucket was out looking for whales in his small boat with the beach of his island home kept well in sight. Suddenly a strong northerly wind whipped up and drove the little vessel far out to sea. While out in the ocean's depths and considerably worried about getting safely to shore again, Capt. Hussey found a great whale and was successful in capturing him. The leviathan was not the usual Baleen or Right Whale to which New England was accustomed, and from which it secured whale-oil and whalebone from the long slabs in its mouth. This fellow was the toothed whale, or, as the French called him—the *Cacholot,* or Sperm Whale, whose oil was very valuable. As it happened, Capt. Hussey's chance catch was the first Sperm Whale taken by an American fisherman. I do not think the feat was achieved by the Nantucket captain alone. He no doubt had the usual complement of men along with him.

The second important result of this accidental excursion was the improvement in American whaleboats which had previously engaged only in coastal whaling. After this, Americans began to build their heavier vessels. Three years after Capt. Hussey's discovery of the Sperm Whale, Nantucket owned six sloops which were engaged in this fishery. In 1725 there were twenty-five American whalers of from 30 to 50 tons burden which pursued whales in deep waters. By this period schooners were introduced for whale catching and the size increased to 70 tons. Mr. Upham intimates that the inhabitants of Salem were not engaged in deep-sea whaling until about 1726. Provincetown on Cape Cod became active in this pursuit by 1737.

In 1804, the year after Salem's Ship *Rachel* sailed on her sealing and whaling voyage to the Antarctic, there began the long era of American whaling which endured with vigor for a period of 72 years, a man's normal lifetime, before its decline. The highest point of America's greatest whaling activity from 1804 to 1876 was reached in 1845 when the importation of sperm oil, whale-oil and whalebone came to well over nine and a quarter million dollars. In 1804 four American seaports dispatched whalers to distant oceans. New Bedford was responsible for 23 whalers; Nantucket for 20; Sag Harbor dispatched 3 and Hudson, New York, 2 whalers.

By 1804, the threshold year of American deep-sea whaling, Salem, now a town of some 9,000 inhabitants or more, was a shipping center of recognized importance. Salem vessels traded in Sumatra, China, the Philippines, in Bengal and Bombay of India, and in Batavia of Java in the East Indies. Two prominent Salem merchants, William Gray, and Elias Hasket Derby, had subscribed $10,000 each for the building of the U.S. Frigate *Essex* for Congress, and the subscriptions of other Salem citizens made possible the building of the frigate which was undertaken by Salem's outstanding shipbuilder Enos Briggs. The East India Marine Society, the present Peabody Museum, had begun in 1799 its vast collections from many exotic lands with a few objects brought from Sumatra by Capt. Jonathan Carnes' vessel the *Rajah.*

In the realm of politics, New England had in 1800 come out unanimously for John Adams as an expression of gratitude to the Federalist Party which had brought the region prosperity. Within Salem itself will be recalled the bitterness between Federalists and Jeffersonians and in particular the feud between the ship-owning families of the Crowninshields and Derbys over the matters of wharf extensions and bridge construction—in the midst of which domestic unpleasantness the Federalist leader, Alexander Hamilton, arrived and was royally entertained by his Salem supporters in the hall built by Samuel McIntire and afterwards called Hamilton Hall.

In the opening years of the nineteenth century Salem vessels struggled with pirates on the Barbary Coasts of northern Africa, and Dr. Nathaniel Bowditch of Salem published his *Practical Navigator,* the mariner's guide which did as much for navigation as Marco Polo's gift of the compass. In 1803 the construction of the Salem Turnpike brought Boston and Salem closer together and contributed to the commercial prosperity of both seaports, while in the same year, the young nation expanded as President Thomas Jefferson purchased from France the Louisiana Territory. And in 1804—the year of Nathaniel Hawthorne's birth on Salem's Union Street—Salem had in active service on many seas 54 ships, 18 barks, 72 brigs and 86 schooners with 48 vessels 'round Cape Horn. Banks and insurance companies were organized to protect precious cargoes in the bottoms of vessels which sailed on every sea. Salem shipping survived the tragedy of the Embargo of 1809, the War of 1812 which followed, and proceeded to open up trade with Zanzibar.

Salem, on the whole, was an aggressive, progressive and busy seaport which had developed out of Puritan beginnings, and it was not unlikely that by the 1830's some of the town's young blood were willing to enter a branch of maritime service considered the most hazardous and daring of all—hunting the sea's largest and most wealth-producing creature. Nor is it surprising that Salem merchants should dare to risk their capital in the whaling industry, *now* regarded in retrospect as the most picturesque that America has ever known, *then* a risky business. All about Salem was the tangible evidence of what success at sea could mean to a man.

As the wealth of the Indies and Cathay flowed into Salem's warehouses, Chestnut Street, laid out in 1796, and Essex Street, too, became lined with imposing homes and gardens of successful sea captains and wealthy merchants. Fine mantel pieces and fence posts of fragile beauty were created by Samuel McIntire for his discriminating patrons. And, in 1782, this Salem carver, destined for renown, had erected on Federal Street his architectural masterpiece, the Peirce-Nichols House, which secured for him assignments to design other elegant homes such as the one now standing on lower Essex Street built for Captain John Gardner and known after 1834 as the Pingree House. Furnished by such masters as Sheraton, Chippendale, and Hepplewhite, this David Pingree mansion epitomized the prosperity of Salem's shipping era at the opening of the nineteenth century.

Only a few seventeenth century houses remain in Salem of today, but a hundred and fifty years ago there were many of these unpainted, wooden structures of the Puritan days, sharp-gabled of medieval imprint, with what Hawthorne called their *beetle-browed second story* projecting over the foundations. A little-remembered seventeenth century mansion was the many-gabled Great House of Philip English, the prosperous merchant who in 1692 owned 21 vessels besides a wharf and warehouses. Such wealth attracted to himself the envy of some of the Salem townsmen who accused the merchant and his wife of witchcraft. The indignities the Philip English family suffered during Salem's period of madness is part of another story. Their old mansion, built in 1683, was familiar to generations of Salem folk. A reminder of Elizabethan England, it was loved by Nathaniel Hawthorne in his youth and somehow it managed to survive deterioration and destruction until Salem's whaling period was well begun.

Long tenantless and memory-haunted, the unpainted house stood on the corner of Essex and English Streets until it was demolished in 1833. The year before it was removed, four whalers sailed out of Salem Harbor, the three full-rigged Ships *Bengal, Catherine* and *Izette,* and the Bark *Pallas,* under Captain Henry Archer, Jr. The *Bengal, Catherine* and *Pallas* went around Cape Horn to the Pacific whaling grounds, the *Izette* to the South Atlantic. When these whalers returned from their distant adventures, laden with greasy spoils, the familiar landmark of the brooding English House had vanished forever.

CHAPTER

4

Salem and the Fur Seal Fishery

"Whatever enlarges hope will also exalt courage."
Samuel Johnson (1709-84)

Early whaling voyages included the taking of seals whose furs could be traded for profit or in exchange for needed whaling equipment. Later specialization set in and you either went for whales or for seals. When Captain Frost maneuvered the *General Knox,* a Salem vessel, safely into the harbor of Boston in 1784, he had in her bottom a cargo of 600 barrels of whale-oil and seal skins taken at the Falkland Islands. Captain Frost's vessel was the second American ship to secure seal furs from those bleak, wind-swept outposts in the Atlantic Ocean, bordering on the Strait of Magellan off Argentina. First discovered in 1592, nobody knows how long the Falklands served as the favorite haunts for thousands of seals, whales and sea-elephants. After 1592, the islands were lost to ken time and again, colonized, claimed and disputed by nations seeking valuable whale-oil and a needed base for fuel and provisions. In 1765, the Falklands were first colonized by a group of Frenchmen fleeing from Acadia. In the hands of the British in 1771, their proprietorship was challenged by the Spanish. Since 1833, this ocean oasis has been consistently held by Great Britain.

Early Salem whalers customarily secured whale-oil, fur seals, and sea-elephant oil, scarcely inferior to that of the whale. Some vessels, however, specifically voyaged for fur seals. The

Minerva left Salem for sealing in the Antarctic in 1799, returning in 1802, and the *Rachel* sailed in 1803 on the same mission. Early seal hunters hoped that the commerce in seal skins would become as lucrative as the sea otter trade or at least that sealing would be a profitable by-line of the whaling industry. Salem's whaling as such did not begin until the voyage of the Ship *Bengal* in 1832, whose sole object was whale-oil.

Seals were brought to the attention of traders by Captain James Cook, the British explorer who was on his third circumnagivation of the globe when he was killed in Hawaii in 1799. He was stabbed to death in the surf at Kealakekua Bay by a Hawaiian, which ended his great seafaring career. Although he was a Britisher and flew the enemy Jack during the American Revolution, Benjamin Franklin saw to it that American privateers did not molest Cook's scientific expedition. Much interest attaches to Cook's voyages and we find men connected with him who commanded attention on their own. There was, for one, William Bligh, the master of Cook's ship *Resolution* who gained fame through the mutiny on his ship the *Bounty*. The death of Captain Cook was avenged by the bombardment and destruction of a Hawaiian village. One of the simplicities of the natives was evident in that the survivors enormously enjoyed the spectacle of the very fires and crackling flames which destroyed their homes and took the lives of their friends and families.

Captain Cook's accounts specifically mention the large seal herds of the South Georgia Islands in the Antarctic. Hardy Scottish fur seekers pushed west across the top of North America. Traders from Russia ranged south from the Aleutians in quest of the rich furs. By one of those oddities in which history abounds, John Ledyard, a Yankee from Connecticut, was the only American to accompany Captain Cook. He stood beside him trying to protect the Captain when Cook was killed. Tall, blonde, grey-eyed, he had been a Corporal of Royal Marines when he joined Cook's expedition.

In the summer of 1785, Ledyard called upon Thomas Jefferson in Paris when Jefferson was the newly appointed United States Minister to France, replacing the aging Benjamin Franklin. Ledyard revealed to Jefferson his story of adventures and his plans to open up (with the help of John Paul Jones) America's virginal Pacific Northwest. Captain Jones was then in France raising money for an expeditionary vessel. At a time when western United States migration stopped at the Mississippi River, the scheme was to land at Vancouver Island to establish a fur post, from which trade could be conducted with the Orient, Europe and the eastern seaboard of America.

Sailing around Cape Horn and the Cape of Good Hope, Ledyard was the first white man to land at Vancouver while he was with Cook's expedition. There he saw thousands of sea otters frolicking in the water. Their pelts were the most valuable in the world. In reporting Captain Cook's adventures and following through with his own, Ledyard can be credited with initiating the profitable seal trade with China.

While in Paris to arrange for his voyage, Jefferson was so impressed with Ledyard, he introduced him to the Marquis de Lafayette who promised to help him. Ledyard was hardly an idle dreamer. He had crossed the hazardous bogs of Unalaska Island to scout a Russian fur post for Captain Cook. The fur traders were so impressed with his heroic proportions, earnestness, and great personal charm they replaced his ragged clothing with a silk shirt, good underwear, a sable cape and fed him roast whale and smoked bear.

Jefferson called Ledyard "a man of genius, some science and of fearless courage and enterprise." And he never could forget him. Long after his retirement from the White House to his gracious home in Monticello, Virginia, Jefferson wrote a book about his young friend who, still adventuring, died at the age of thirty-eight, having become so enraged at some Arabs he broke a blood vessel in his temple.

There are many tales and few to tell them. Before Captain

Frost engaged in sealing at the Falklands, we learn that a wealthy Boston woman became indirectly involved in this pursuit, which was highly profitable. Shortly after the Revolution, Lady . . . . Haley arranged for the outfitting of the Ship *States* of 1,000 tons burden. The object was the securing of sea-elephant oil, however when the Captain of the *States* arrived at his destination and saw literally acres of richly furred seals basking lazily in the sun on the beaches of the Falklands, he by-passed his original instructions to take sea-elephant oil and set his crew instead to work to take seals. The *States* is the first known American ship to have taken fur seals. With this voyage begins the exciting story of the southern seal fishery.

The *States* remained at the Falklands for a period of two years. In that time 13,000 seals were killed and their skins removed. It was hoped that the long voyage and months of gruelling work in the desolate islands would bring a fitting reward. When the *States* returned to Boston laden with its promising cargo, the reception was lukewarm. Domestic furs were in ample supply along the Atlantic coast and much labor would be required to prepare the 13,000 skins for use. Finally, they were sold at a low price to a New York merchant, who, it is said, believed he was purchasing sea-otter skins. All he would pay was 50¢ a piece, so the complete cargo became his own for $6,500. From this point on, the furs were headed for a brighter destiny.

Whatever his private beliefs on the identity of the furs, the New Yorker sold them to a Calcutta concern as sea-otters. From Calcutta they were brought to China by Captain Metcalf of New York. In Canton, the fur seals secured by the Boston woman's investment sold for $5.00 a skin, or ten times the original selling price. The $65,000 transaction caused a stir among the American traders doing business in China. In 1789, the year of the sale, American-Chinese trade was brisk. Fifteen American vessels are known to have been at Canton. Five flew the commercial flags of Salem firms. All but one of these vessels were owned by Elias Hasket Derby, the affluent and brilliant

Salem merchant. The Americans in China were quick to recognize the furs as a new medium of exchange to pay for the teas, silks and nankeen (cloth) wanted from the Orient for the American market. Encouraged by the needs of American trading captains, the fur seal industry came into its own.

In the very early ventures, Salem took no direct part in sealing. The Boston Ships *Hope* and *Jefferson* made voyages to the seal rookeries on their way to the North West Coast before 1794. During the years 1794 and 1795, there appears to be no record of any sealing voyage from an American port for furs to be traded in Canton, where a break in the market made furs unprofitable. During the quiet period, shipments of seal furs continued to find their way to the Chinese market. Whalers took numbers of seals during their whaling quests and when the oil was unloaded in New England ports, the furs became part of the cargoes destined for the Orient.

Yankee traders were the first to begin the seal fur trade with China, the seals making their entry under the guise of the already acceptable sea-otters. Some twenty years before Captain Cook's explorations which led to the discovery of rich seal rookeries, the Russians had sent North Pacific sea-otters into China, shipping them from Siberia by way of the River Kamchatka. Not until 1786 were the fur-bearing seals found in such tremendous numbers along the coasts of Alaska in the Bering Sea. Their discovery was made by the adventurer Vitus Behring, the young Dane employed by Empress Catherine of Russia. When the abundant seal herds of the Antarctic were reduced to pitiful remnants, the Pribilof Islands in the North Pacific became the hunting grounds. Named for their Russian discoverer, Gerassim Pribilof, they are more commonly called the Aleutian Islands, which were created a fur-seal reservation when purchased with Alaska in 1867 by the United States.

After the decline of the Chinese market, the sealing industry was greatly encouraged by the remarkable performance of the *Neptune,* a vessel of New Haven, Connecticut. The *Neptune*'s success became one of the legends in sealing annals. She was owned almost wholly by Ebenezer Townsend whose son, Ebenezer, was supercargo or business manager. Ebenezer Townsend, Jr. kept a careful, intelligent journal of the voyage which throws considerable light on a rather obscure phase of our maritime history.

The Neptune left New Haven in 1796 and returned in 1799. The voyage had been organized on an ambitious scale and produced 80,000 seal skins, which sold in Canton at $3.50 each, or a total of $280,000. In exchange for the value of the furs, the *Neptune* imported from China a fabulous, assorted cargo of teas, silks, jades, teakwood, etc. for both American and European markets. The amount of custom duties paid by the *Neptune* on this voyage gives some idea of the magnitude of the trade for which the seal furs served as a medium. The duties came to $75,000 which exceeded the Civil List Tax of the entire state of Connecticut.

Many in Salem were not unaware of the seal herds and their profitable possibilities. The year the *Neptune* began the voyage which culminated so brilliantly, George Nichols, a Salem shipmaster and merchant, sailed from Salem on May 10, 1796 in the Brig *Eunice* on a commercial voyage. On his return, Capt. Nichols reported vast seal herds on the islands of St. Paul and Amsterdam, two or three degrees south of the Cape of Good Hope, Africa. It took the *Eunice* four and a half months to reach the Cape and the islands lay about 20 days farther on. "Knowing that seals and fish were very abundant at St. Paul and Amsterdam" Nichols wrote in his *Autobiography* "we concluded to land at St. Paul. The whole shore was lined with seals . . . We encountered as we went an incredible number of them. In one small place there appeared to be more than one hundred lying basking in the sun asleep like so many sheep. We succeeded in knocking down a great many."

It is not surprising that the alert merchants of Salem should turn to the seal fisheries to secure a welcome medium of exchange for coveted Chinese merchandise. The year the *Neptune* returned laden with exotic spoils in 1799, Salem entered the Ships *Concord* and *Minerva* into the seal gold rush. A great deal of information is available from the log of the *Concord* which was kept by Nathaniel Appleton, who for a time was transferred to service on the Salem vessel, *Minerva.* For many years Nathaniel's father, John Appleton, Harvard 1757, was a prominent importer in Salem. Nathaniel's mother was Jane Sparhawk, daughter of John Sparhawk, who became minister of Salem's First Church in 1738. John and Jane Appleton were an integral part of Salem's cultural life. It would hardly seem they had time for there were fifteen children of whom Nathaniel, our logkeeper of the *Concord,* was the sixth.

At the time of the *Concord*'s sailing, Nathaniel was twenty years of age. The *Concord,* 171 tons, Capt. Wyer, was owned by several Salem merchants: Pickering Dodge, John Barstow, Gideon Tucker, Benjamin Dodge, Benjamin Dodge, Jr. and Nathaniel Appleton. As Nathaniel appears as an owner, it is probable that John Appleton presented his son with a share in the *Concord* after Nathaniel's satisfactory service in his father's "counting house," or business office. From the well-kept journal of the *Concord* it is apparent that Nathaniel Appleton was conscientious and well-trained. This interesting record is now in the marine files of The Essex Institute.

As for the *Minerva,* 224 tons, built in Haverhill, it was owned by Nathaniel West and Timothy Williams, and commanded by Mayhew Folger. For many years Capt. Folger had been engaged in the fur seal fishery. Several years after this voyage, he became connected with one of the fascinating legends of the sea for which he is more apt to be remembered than for his sealing talents. While commanding the *Topaz* owned by Thomas H. Perkins of Boston in 1808, Capt. Folger searched the seas for new sealing grounds, and in this search happened upon Pitcairn Island where he found the single survivor of the famous mutiny on the H.M.S. *Bounty,* Captain William Bligh.

When the *Concord* and *Minerva* sailed out of Salem Harbor in October of 1799, the owners could reasonably hope for success. Other sealing voyages had proven highly profitable. Capt. Edmund Fanning of New York had already profited greatly from taking seal skins. While commanding the *Betsey* of Stonington, Connecticut, in a two-year sealing voyage, Capt. Fanning discovered Washington Island and Fanning Island in the South Pacific. In addition to the islands, Capt. Fanning chalked up $120,000 from the cargo with which he returned under the *Betsey*'s hatches.

For many weeks the sea-journal of the *Concord* contained only weather entries and nautical details. Christmas came and went without comment, a left-over of Puritan times when Christmas was not celebrated. New Year's Eve received some recognition by young Appleton. On the last day of the old year and the vanishing century, he wrote: "The sun was exactly vertical today at 12 meridian. Farewell to the year 1799. I feel some regret in leaving you, thinking of the many happy hours I have passed in it."

By January 27, 1800, the *Concord* had arrived at Grand Jason in the Falklands. Here they found the Ships *Barclay,* Captain Barney of New Bedford and the *Mars,* Captain Swain of Nantucket, both bound 'round the Horn a-skinning. All through the sealing and whaling years, wrote Appleton, New Islands was a famous place of rendezvousing and refreshment with fresh water plentiful and ship fuel. Trees there were none, but in lieu of firewood, the barren islands provided an excellent quality of peat.

Appleton's minute descriptions of birdlife are of particular interest to ornithologists. One of his entries tells of going on shore and seeing "a great many birds such as shags, albatrosses, geese . . . the albatrosses and penguins were so thick that for several acres you could not walk without treading on them. The penguins look like a large army of soldiers on parade." Appleton tells also of the shrieking sea-birds which afforded good eggs in boundless quantities. When Edmund Fanning stopped at New Islands, he laid down 56 barrels of these eggs for sea stock during his odyssey in the Pacific. The islands abounded, too, in wild cattle and hogs. These were the progeny, says Appleton, of breeding animals which had been abandoned by early colonists who left the islands forever. The homeless cattle fed on tussock-grass peculiar to that region. The tender asparagus-like shoots of the tussock-grass were a delicacy relished by the *Concord*'s crew separated for four months from garden greens.

Fish there were in endless tasty varieties. Among Edmund Fanning's numerous contributions was his discovery in 1801 of cod fish in this section of the globe. Mussels and clams were there for the digging. The day after the arrival of the *Concord* in the Falklands, the journal records a social visit paid by the Captains of the *Mars* and *Barclay*. Captain Wyer's invitation to board the *Concord* extended to Captains Swain and Barney was for the evening meal. It was a good opportunity to compare notes, swap news and consume a little rum. Perhaps the most important bit of information Capt. Wyer secured from his dinner guests was that the year before they had been at Canton with 21,000 seal skins taken at Masafuero.

After tarrying at the Falklands for three weeks and a day, the *Concord* got under way on February 18th in company with the *Mars* and the *Barclay*. The course was laid for Cape Horn by way of the straits of Le Mare. Getting around treacherous Cape Horn was a perilous undertaking. Appleton complained bitterly of "plagy, dirty, good-for-nothing weather" destructive to the masts and rigging. "I shall be an amphibious animal" he wrote in his log "before I get home for I live as much under water as over it."

Gales and lashing rainstorms, punctuated by snow and sleet, threatened to terminate the *Concord*'s sealing adventure minus profit or glory. "At 8 P.M." wrote Nathaniel Appleton "there was carried away our main topmast, top gallantmast, and fore topgallant yard, owing to the main top's giving way and slackening the rigging. There was our main topmast and top gallantmast with the sails in the water together, besides a lee shore on which we expected to drift every moment. If this be Cape Horn, spare me from ever coming again!".

Captain and crew were grateful when the winds died down sufficiently to allow for repairing the damaged masts and sails. The *Concord* was again set in her course around the troublesome Horn, but now the winds failed to rise sufficiently to fill the sails. The ship remained quite stationary. "The plagy ship" mourned Appleton "will not go better than North on one tack and South on the other, so there is no chance of our ever getting round into the Pacific Ocean. The wind is no more likely to alter than the sky's falling."

Appleton was almost right in his gloomy prediction. The next day brought calm again. It prevailed for 24 hours. Then came piercing head gales and "thick, dirty, cold, disagreeable weather" which held on for "a long spell." After four days of it, a Friday at last brought "strong gales from the west, rain, hail, snow and blow and people got all their clothes wet and had no dry ones to put on." Appleton began to feel sorry even for himself. "I, a poor fellow" he wrote descriptively "sitting down in the cabin with all deadlights lashed in, writing here with a nasty lamp which fetches way every minute and oils me all over. So ends this day."

Finally, the *Concord* managed to get around Cape Horn for on March 16th there is an entry in the log which tells of their anchoring at St. Mary's on the Chile Coast. St. Mary's played the same part as a whaling and sealing rendezvous in the Pacific as the Falklands played on the Atlantic side of South America. St. Mary's was an important seal rookery and preparations were made to send the skinning gangs ashore. For several weeks before reaching the sealing grounds, the crew were occupied with preparing the tools and equipment needed in taking the seals and in removing their skins. The weapon used was a club made of hickory which resembled a baseball bat. A single blow was enough to crush the seal's light skull bones. At St. Mary's the results were not particularly productive so the *Concord* pulled out for Masafuero where the Captains of the *Mars* and *Barclay* had reported such satisfactory results. Upon their arrival, there was ample evidence that the sealing was good. Eight other vessels were there on the same errand. A rather large number of men were employed in dispatching seals, and there were the usual discontent and quarrels wherever work is being done under pressure and the hopes of reward are high.

At this time, wrote Appleton, there were at Masafuero 14 "lone men" who had deserted from whalers or sealers. They sustained themselves by swapping skins with visiting sealers for food and clothes if they could get them. The *Concord* log tells of the desertion of two of its men at this lonely outpost. "In the course of the night Glover and Drowne, two of our sealers, stole the yawl and ran on shore with all their clothes. We found the boat they used, but cannot find the men." Three days later . . . "Glover, the fellow that ran away, came back and said he was very sorry. Caught Drowne, but Moser kept his distance. Night calm. Some hands on shore to catch the villain. Morning calm, but no Moser to be found. The fellow must be a plagy fool, for he's got no clothes but what he has on, no fire-works, no nothing of the kind."

During the sealing season at Masafuero about 200 men were employed in taking seal skins. The number of seals killed was very high. When Capt. Fanning left for Canton two years earlier, he had 100,000 skins on board the *Betsey*. It was his belief that some 750,000 seals remained on the beaches. Benjamin Morrell estimated that by 1807, three and a half million seals had been killed. The crew of the *Concord* took between 8,000 and 9,000 skins in three months' work. The *Concord* remained at Masafuero until August of 1801, which was almost a year and a half after her arrival. If the men continued to take at the same rate as for the first three months, the cargo of the *Concord* must have been close to 50,000 skins. On August 28, 1801, the *Concord* sailed from the sealing grounds at Masafuero for Canton by way of the Marquesas Islands and the Island of Tinian. In Canton, Captain Wyer disposed of his furs for an exchange cargo and began the long trek home to Salem. The entire voyage lasted from October 1799 to July 1802. The expense of the voyage, the crew's participation and the outfitting of the vessel was set down as $25,142. On July 28, 1802, the *Concord* and her cargo sold at Salem for $67,794.56.

The *Minerva,* which departed with the *Concord,* returned to Salem ahead of her, docking on May 4, 1802 according to the *Salem Gazette*. Without any supporting detail, the *Gazette* reported: "May 4, 1802. Arrived Ship *Minerva* Captain Folger, 153 days from Canton. Sailed from this port on a sealing voyage October 1799, and has been successful." In 1803 the Salem Ship *Rebecca* made a sealing voyage to Masafuero, and the Ship *Rachel* went sealing to the Antarctic the same year. There follows a number of years in which no Salem vessel appears on the records of the trade. The southern seal fishery had entered a long decline.

Sealers were indiscriminate in their destruction. The male seal matures at seven years of age when he can reach a weight of about 500 pounds. This type of larger, stronger male dominates by his physical superiority. Like the schoolmaster bull of the whale family, a big seal heads a family or colony of some 25 to 30 females and their cubs. Each female bears a single cub at a time. The lighter males are free of such domestic responsibility as some of their more heroically proportioned brothers seem glad to assume. The happy bachelors travel by themselves. It is these herds of young males, which under controlled sealing, supply most of the skins of commerce. In the 19th-century-haste to kill anything covered with a furred skin, even the family heads were not spared to assure the perpetuation of the race.

The South American herds were practically annihilated, as well as those frequenting the Australasian Islands. It was inevitable that sealing should become unprofitable.

Some fifteen years after the voyages of the *Rebecca* and the *Rachel,* enough seals reappeared to allow the industry to revive. We find the Brig *Britannia* of Salem being outfitted in 1818 for a sealing voyage which never materialized. A few hours after leaving Derby Wharf, the *Britannia* was wrecked off Manchester, Massachusetts. Two years after this disaster, the numerous herds of the South Shetlands in the Antarctic were discovered. Again sealing was eagerly undertaken by Americans. No less than 17 American ships hurried to the rocky Shetland coasts to help themselves to thousands of skins. In this sealing fleet, Salem was represented by the *General Knox,* commanded by Captain Orne. After a year, the *General Knox* returned to Salem with a cargo of 5,000 skins and 600 barrels of sea-elephant oil. In 1821, the Salem Brig *Nancy* made a voyage for the purpose of securing whale-oil or seal skins. She returned with neither, but hailed instead 100 barrels of sea-elephant oil. This rather unfruitful voyage of the *Nancy* seems to have been the last attempt of this seaport to secure fur seals.

It is possible that by the time the *Nancy* arrived at the sealing grounds, there was nothing worth taking. So rapid was the destruction of seals at the Shetland Islands that within two years of their discovery, the large herds had been reduced to a few stragglers. It is true that considerable wealth was brought into the United States at the time of the extensive sealing raids when currency was scarce, but it is doubtful that the nation's economy benefitted sufficiently to offset the loss of these herds to posterity.

CHAPTER

5

Whale-oil, Whalebone, Whale Beef

"To business that we love, we rise betimes."
Shakespeare

Whaling was complicated in technique, hazardous not only to life and limb but to the pocketbook as well, required infinite patience in cruising for weeks and months on end in the search, and the combination of various skills, yet men built whalers, outfitted them, and other men took them whaling in distant, dangerous waters. To understand why they did so requires some understanding of the uses for the products which the whale offered, which made his pursuit and capture so worthwhile.

Before the era of petroleum, whale-oil served many purposes, chief of which was illumination. Capt. Scoresby gives a full account of its uses up to 1820. Other uses were later evolved. Whale-oil once lighted most streets of European towns. In America, whale-oil was used for street lighting, for the powerful beacons of lighthouses along the coast line and as headlights of locomotives. There was a great variety of lamps for home illumination. The use of lamps and machinery increased during the early nineteenth century and caused a greater demand for sperm oil. The very general adoption of gas for purposes of illumination which might reasonably be expected to lessen demand had the opposite effect. The light of shops and street induced persons to brighten their households in a proportionate degree. However, the popularity of whale-oil as an illuminant began to decline around 1840. When the price of sperm and whale-oil put it out of reach of many home owners, some alert Yankees began to make in quantities lamps suitable for burning lard oil. The increased use of gas and electric lighting during the nineteenth century modified the use of sperm oil, but it remained important as a lubricant.

Sperm oil has a peculiar property which enables it to withstand great heat and cold, and makes it adaptable as a lubricant for precision machinery. Sewing machine oil, for instance, was a combination of sperm oil and bone-filtered petroleum. The sperm oil was refined by a thorough process of heating, freezing, pressing and bleaching. Soap made from sperm oil was used in cleaning harnesses and saddles, and spermaceti was reserved for making the finest candles. Whale-oil, extracted from certain species of Baleen or whalebone whales, was also used for a lubricant and served in soap manufacture. There are five grades ranging from No. 0, the lightest and best, to No. 4, the darkest. Color and grade depend, of course, upon the part of the whale from which it is extracted and the care with which it is refined.

When one considers that the average barrel held 30 gallons of oil, and whalers returned with cargoes which sometimes ran to 2,000 barrels, the value can readily be appreciated. When the Ship *Elizabeth* of Salem returned from the Pacific in 1844, she hailed 1,500 barrels of sperm oil, at a time when sperm was worth 88¢ a gallon. Roughly, the sperm catch was worth $39,600.00. The *Elizabeth* reported no other cargo, so either fished strictly on a basis of "sperm or nothing," or obtained whale-oil and whalebone and disposed of them before coming home. The *James Maury* reported back to Salem in 1845 after a four-year stretch with a cargo of 1,400 barrels of sperm valued at $36,960., 500 barrels of whale-oil, which at 33¢ a gallon were worth $4,950, and 3,600 pounds of whalebone which at 34¢ a pound was worth $1,224, or a total value of $43,134 less custom duties.

Prices for whale products varied with scarcity, demand, rivalry of other substitutes, and trend of the times. A glance at Starbuck's account of prices for the American whaling years from inception to 1876 gives some idea of the values involved which kept whaling alive. In the early period, sperm is listed at 93½¢ a gallon and whale-oil at 35¢ for the year 1820. Ten years later, sperm had gone down to 65½¢ and whale-oil was still 39¢ a gallon during the 1830's. In 1837, which happened to be the year in which Salem sent out her greatest number of whalers, sperm was 82½¢ a gallon. Incidentally, this year was the biggest sperm year of all in American whaling. The total reached was 5,319,138 gallons secured for all ports.

In 1839 sperm rose to $1.05 a gallon. During this year, the Bark *Derby* returned to Salem with 180 barrels of sperm worth $5,670. The *James Maury* returned in December of 1839 with 250 barrels worth $7,875, and the little schooner *Mac* reported back in the spring of 1839 with 500 barrels of sperm worth $15,750 plus other cargo. The Bark *Richard* hailed the same number of sperm barrels worth $15,750. All whalers carried large quantities of whale-oil and bone. The *Maury* had 2,600 barrels of whale-oil in addition to the sperm and the *Mac* had 1,000 barrels of whale-oil.

Between 1841 and 1846 the price for sperm hovered around 90¢ and in 1847, it went back to $1.00. In 1850 it climbed to $1.20 and kept rising until in 1855 sperm oil was worth $1.77 a gallon and stayed there up to the Civil War. With the whaling disasters of the Civil War period, the decline of whales caught, the price for sperm rose to $1.89 in 1864 and the next year went to $2.25 a gallon. It reached a high in 1866 of $2.55 a gallon. From then on it decreased until in the last year of the Starbuck record, 1876, it stood at $1.40½. In 1845 whale-oil was 33¢ a gallon and went from 71¢ in 1855 to $1.28 in 1864, reaching $1.45 in 1865 and then declined steadily to 59¢ in 1876.

Whalebone in 1837 was 20¢ a pound. In 1845 it was 34¢ a pound. In 1852 it went up to 50-5/6¢; in 1857 it stood at 96¾¢ a pound; in 1863 at $1.53 a pound; 1864 $1.80 and up to 1869, it was above $1.00. In 1872 it was at $1.28 and finally rose to $1.96 a pound in 1876.

The price of whalebone increased as uses for it were developed. Around 1850 it was much in demand and brought good prices. This springy, flexible, elastic, strong, horny substance came principally from the Right Whale, and was also found in the Bowhead, Sulphur-bottom, Humpback and California Grey. It occurs in the mouth of the Right Whale in long, thin plates, or black slabs of bone fringed at the edges, varying from ten to nineteen feet in length. When the bone is the maximum length, its complete weight is close to 3,000 pounds. Some 300 and more of the horny slabs hang from both sides of the upper jaw and act as a strainer to detain the whale's food when he ejects the water he has swallowed. This most convenient mechanism to the whale, which Melville calls his "Venetian blinds," proved highly useful to mankind.

Whalebone lent itself admirably to the construction of horse whips, an important item when the horse and buggy was the chief means of transportation. It is difficult to convince anybody in New England that their grandfathers used anything but a genuine whalebone whip. A man inquired at a store for a whalebone whip with masculine pride in quality, and the Yankee salesman talked about whalebone with the awe applied to *all wool* and *pure silk*. It was the best thing on the market. In Westfield, Massachusetts, the whip industry which originally paid 25¢ a pound for whalebone had to pay as much as $10 a pound for it at the peak of whip production in 1910, just about the time the automobile was replacing the horse. Finally the price of whalebone rose to $100 a bundle as whaling disappeared from New England ports. Then Westfield manufacturers began to manufacture raw-hide whips.

Whalebone was used in brush making early in the nineteenth century by boiling it until it was soft and splitting it very fine. It was either used alone or mixed with bristles as a cheaper but less durable substitute. The splitting was done "with a number of lances fixed in a handle." Whalebone was popular also for making ribs of umbrellas and for parasols, for surgical instruments, for sofa and carriage springs, and for stiffening clothing in an era when women were addicted to bustles and hooped skirts.

Many a beautifully attired woman danced in gay ballrooms lighted by sperm candles in gowns which patient seamstresses had supported with whalebone. Beneath the outer glory was the reliable and supple whalebone corset. Towards the end of the nineteenth century, the whale fishery depended almost exclusively on the sale of whalebone to corset manufacturers as the wasp-like 19-inch waist and tight lacing confined feminine flesh to the narrower and more attractive restrictions demanded by Dame Fashion. By 1906 whalebone had risen to $10,000 a ton, or $5.00 a pound. The hour-glass figure vanished although less drastic corsets remained. Less satisfying substitutes at last replaced the now too-expensive whalebone corsets. In New York alone there were eight bone-cutting houses when the Pacific Steam Whaling Company was one of the principal sources of whalebone. An old shop at 161 Duane Street, New York City, kept by George Messmann once handled 30,000 pounds of whalebone a year. The last of New York's noted whalebone dealers, Mr. Messmann closed his doors in September of 1920. By 1927 the once prized whalebone had become a drug on the market.

Another valuable product from the whale was the rare ambergris which brought from $200 to $400 and more for a single pound. Once mistakenly used medicinally, ambergris is prized as a fixative of the fragrance in expensive perfumes. No Salem whaler was fortunate enough to enrich its cargo with ambergris, but every skipper and man in the crew searched excitedly for it in the carcass of the whale after the blubber had been removed. A rich find of ambergris might be worth more than all the barrels of sperm and whale-oil in the hold of the ship.

The Arabians were the first people to prize the waxlike substance found in the intestines of the Sperm Whale and they called it *amber*. Sometimes ambergris has been picked up near the whaling grounds as it floated on the surface of the sea. Early travelers along the shores of the Indian Ocean and to Southern Asia frequently referred to ambergris in their writings. Even before the time of Marco Polo who was born in 1254 and died in 1324, Zanzibar was famous for its ambergris. There are islands and mountains in Zanzibar called by this name because of the precious finds there in the sixteenth and seventeenth centuries.

Until recent years, the rare substance has been shrouded in mystery and even fairly up-to-date dictionaries do not give a satisfactory explanation. "It is now generally agreed," according to Mr. William Tripp, the curator of the New Bedford Whaling Museum, "that ambergris appears in both sexes of the Sperm Whale, but more frequently in the male. It is found only in an ailing whale, great lumps having formed in the distressed mammal's intestines as a result of indigestion." Ambergris looks like cork or a rubber sponge. It is ashen in color and forms in chunks which weigh from a pound to 200 pounds or more. Ambergris has a naturally sweet aroma and is often used in the sacred rites of the church. In early days, it was mistakenly valued as a medicine. In Asia, it was used as a spice in the cookery of the wealthy. Now ambergris has been found to be an ideal fixative for volatile oils and other aromatic fluids of which expensive perfumes are made. The ambergris preserves the fragrance of the perfume for long periods because of its curious property of retaining or absorbing other odors. Inexpensive perfumes contain no ambergris and quickly lose their scent. The smell of ambergris itself is pleasing and has been described as reminiscent of musk or of newly plowed earth. You can easily test its authenticity. Ambergris will dissolve in cold alcohol. If you take a heated needle and pierce your specimen, it will be oily when removed if true ambergris. If a sticky substance adheres to the needle—it is not ambergris.

Ambergris is as eagerly sought as pearls or coral. After the blubber and meat have been cut from the carcass, the whalemen use the cutting-in spade to puncture the intestines. When ambergris is present, there is a rubbery resistance to the spade. A whale which appeared thin or yielded little oil was a likely prospect for the feverish search. When ambergris was struck, the excitement on board ship exceeded anything else in the voyage. Sometimes its value was greater than the entire cargo.

In the hey-day of whaling, fortunes were made in ambergris. There was the famous buyer and wholesaler of Provincetown who dealt in ambergris for half a century and won the title of the Ambergris King. His name was David C. Stull. The largest catch of the precious substance was made by the Bark *Splendid* on a voyage out of Dunedin, New Zealand. From a single whale was taken 983 pounds of ambergris. It took ten years to dispose of it and eventually brought $125,000. In 1921 the New Bedford Schooner *Valkyrie* brought home a 74-pound cargo of ambergris which went to a New York firm for $9,000. No Salem whalers are known to have taken ambergris.

Salem whalers may not have secured ambergris, but they often enough hailed huge cargoes of whalebone. When the Bark *Derby* returned from the Indian Ocean in 1839, she had 8,000 pounds of bone on board. In 1843, the Ship *Emerald* reported 8,800 pounds and in 1845 the Bark *Henry* imported 2,400 pounds. On board the *James Maury* during the same year of return—after four years in the Indian Ocean, were 3,600 pounds of whalebone. In 1848 we find the Ship *Elizabeth* returning to Salem after four years in the Pacific with 14,000 pounds included in her cargo. In 1856 the Bark *Messenger* returned from the Atlantic with 1,200 pounds and in 1866, when whalebone was almost $2 a pound, the *Falcon* brought in 470 pounds. Other amounts of whalebone may have been procured by Salem whalers and traded or sold before the vessel docked in its home port. However, the figures actually imported amount to 38,470 pounds, a not inconsiderable quantity for corset stays, whips, brushes, umbrellas and the many other articles fashioned from whalebone from the whale's mouth.

In the sailing days, whale-oil, sperm and whalebone were the chief prizes—oil to light the cities and baleen for corsets and horsewhips. Now, whale-oil lamps rest idly on museum shelves, but the demand for the oil which once lighted them has, with the last decades, grown among manufacturers of soaps and cosmetics. Tremendous amounts of whale-oil are now imported into the United States by some fourteen companies in Chicago, Baltimore, Boston, Everett and New Bedford, Massachusetts, New York City and San Francisco according to the *Thomas Register.*

In April of 1931, the British Whaler *Sir James Clark Ross* delivered to an American soap manufacturer 55,000 barrels of whale-oil worth between 1½ and 2 million dollars. To produce this great cargo the *Ross* cruised in the Antarctic for eight months and secured 1,444 whales. The barrels were unloaded at a wharf in Staten Island, New York. Mr. C. H. Townsend, the whaling authority of the New York Aquarium, went on board and saw the shipment.

In the early 30's, it looked as if the uncontrolled whale hunt had reached a point of diminishing returns. Too many whales were being killed and the oil became a drug on the market. Mr. Townsend received a letter from a friend in Tonsberg, Norway, in March of 1932 which tells of idle whaleships and idle men:

> "So far as I can gather the outlook is this: Next season only the modern vessels will go out and that only provided they have been able to sell the oil in *advance*. There is still a quantity of say 500,000 barrels unsold of last season's catch . . . Sandefjord, Tonsberg and Larvik are the New Bedford, New London and Nantucket of Norway of the present day with Sandefjord lead-

ing. I visited that place the other day and I must confess it was a magnificent sight. Whale catchers in long rows one alongside the other and the huge factory ships completing the picture. Some 8,000 men are idle at home this year. Only two fleets those of Leith and Liverpool comprising 4 factory steamers with their complete set of whale catchers have been sent out."

During the inter-war period when most whaling ceased, whale-oil gained increased importance in the European market as an Anglo-Dutch research organization developed technical processes for its use in a growing list of products. Now the importance of whale-oil overshadows that of other whale products such as ambergris, sperm oil, whalebone and hormones. In Britain before the war, whale-oil provided 41% of all raw materials in margarine, 28% of those in lard compound and 17% of the oils and fats in soaps. Most of the time whale-oil undersold competing vegetable oils in Britain.

Used generally for lubricants and in soaps, it became possible to use whale-oil in the manufacture of edible products, particularly margarine, when a German chemist eliminated the fishy taste by hydrogenation. From 1926 to 1931 the whale-oil industry enlarged its volume by 300%. This meant that whale-oil was being used instead of lard which represented 11% of American hog products. By 1931 edible whale-oil presented a problem for American meat packers. Whale-oil challenged the place long held by American lard in the European market by amounting to more than half the volume of lard produced in the United States. American packers tried to introduce edible whale-oil into this country but the American public did not take to margarine which used whale-oil as a base.

Things came to such a pass that the Cudahy Packing Company told its stockholders in a report of 1931 that the substitution of whale-oil was responsible for the decrease of sales of lard. *Business Week* of March 11, 1931 protested that Iowa farmers were "faced with a whale of a competitor." To protect the farmers, a federal excise tax of 3¢ a pound was placed on imported whale-oil. American whaling firms such as the Western Operating Corp. which owned the factory ship *Ulysses* and the American Whaling Co. which owned the *Frango* and small units along the Pacific Coast—fought the tax for a while then disposed of their oil quite profitably in foreign markets. As Germany and other nationals were hampered by the war, whale prices in 1939 went to $160 a ton. In 1938 before the federal tax, domestic whalers had supplied the United States demand for 70,664,000 gallons of whale-oil. The whaling companies denied that it competed with any farm product. They warned that if whalers were taxed out of business here, Americans would have to pay more for their soap for which whale-oil acts as a whitener. Through their huge purchases, Procter and Gamble, manufacturers of Ivory soap, dominate this market.

The value of whale-oil is enhanced by its ability to keep over long periods. One cargo of British whale-oil was stored in the United States in 1941 and not made up into margarine until 1947. At the time of its manufacture, it yielded enough to supply the fat ration for the entire British population for five weeks. The oil is graded as soon as it comes out of the pressure boilers on the factory ships. The two top grades of a light golden color are mostly used in comestibles such as margarine, in soaps, cosmetics, candles and for preparing textile fibers for spinning as well as for industrial treating baths. Low grade whale-oil is used as quenching baths for heat-treating steel, for the base of some of the heavier paints and the manufacture of linoleum. The poorest oil, extracted from pressure cooked head and bones of the whale is used in bitumen mixtures for waterproofing and sealing pavements.

The residue of boiled down flesh and bones forms *whale meal* when it is dried, an important part of cattle and chicken feed. The stripped bones once set adrift to make a feast for sharks are now ground up. As *whale guano* it is an excellent fertilizer. Shark livers have long been used as a vitamin source. The extraction of Vitamins A and B from the whale sets a new value on his liver which weighs a ton. The balance is used for feeding fish in hatcheries. From the whale's pancreas is taken insulin for diabetics. The oil of the Blue Whale contains a large percentage of glycerine which is important in the manufacture of explosives. Phosphates are also supplied by the whale. Once so highly esteemed, whalebone no longer enjoys a top place on the list of whale products. As it happens, the whalebone from Humpbacks, Finbacks and Blue Whales has no commercial value except for fertilizer.

Sperm oil, when taken, is used for lubricating precision instruments, watches and for expensive face creams. In 1945, leather from the whale's hide began to be used commercially. The hides of the Belugas or White Whales have long been thus used. Just before the war, the Germans worked on a method of turning whale blubber into artificial wool. The Norwegians tried it but failed. Apparently the process involves putting slices of blubber through special pressing machines and treating the dry residue with secret chemicals which produce thin, strong fibers. War stopped this German experimentation. New products are derived from the whale from time to time, and for many years no part of him has gone to waste. The modern list of derivatives would prove amazing to the whalemen of an earlier century, but this is a scientific age and if George Washington Carver could coax so much from the tiny peanut, is it so surprising that the greatest of mammals should contribute so much of himself to so many different products?

Whaling flourished as one of the backbone industries of American economy from colonial times until after the Civil War. Both the Right or Baleen and the Sperm Whale were sources of many products. The average whale yields from 20 to 60 barrels of oil while the Blue Whale or Sulphur-bottom, a Right Whale, the largest of all species ever known on earth, is a virtual oil well. He reaches a maximum length of 100 feet with a surface area of something less than 2,000 square feet. An 89 foot whale weighs 119 tons and gives 166 barrels of oil. Many thousands of these rich producers were taken off the Norwegian coasts since 1865. Their haunts are in the polar pack ice of both hemispheres.

The blunt head of the Sperm Whale weighs about 30 tons, or a third of his total weight. The upper part of the head which is free from bone is known as the *case;* the lower part of bony structure forming the cranium is called the *junk* and beneath that is a large section of crossed fibers called *white-horse,* forming the upper jaw of the Sperm Whale. The large cavity or sack of the case contains oil in its natural fluid state. Naturalists believe that the Sperm's great tank of forehead oil is a hydrostatic mechanism which helps him in diving to great depths. In its natural fluid state, the oil is colorless but rapidly congeals in the open air. The granulated spermaceti is beautifully clear white. About 500 gallons of liquid sperm, or 15 barrels, can be taken from the case. This head oil, as any other animal fat, has to be heated to prevent it from turning rancid. The junk also yields sweet clear oil which is sometimes equal to its own bulk. When the great head is hoisted up and down against a ship's side after decapitation, it measures about 26 feet.

Only an oil well could compete with production of this volume. In 1859 nature presented this new miracle as the first oil well gushed up in Pennsylvania to declare its superiority. The rivalry between kerosene and whale-oil was sharp and brief. Whereas whale-oil and sperm oil were foremost as lubricants, now lubricating oils were manufactured from the residuum of kerosene. The introduction of wax and paraffin for making candles finally settled the battle against the whale.

In addition to whale-oil, sperm and whalebone, the teeth of the Sperm Whale—the only large whale with dental equipment—were valuable. Weighing about 6 pounds each, the teeth numbered over 40 and were known as whale *ivory,* much coveted for Scrimshaw.

For many centuries the whale has served as an opulent source of meat supply. Whale meat was commonly eaten by the Greenlanders and Icelanders who soaked it in sour whey, and whale beef has long been in use in Japan and Europe. There is a tenderloin near the tail, two large pieces of meat on each side of the backbone of 500 to 600 pounds of red meat of a coarse grain which is sweet and good. It can be ground into hamburgers then mixed with salt pork, seasoned, made into balls and fried, according to the recipe given by John Cook in his *Pursuing the Whale.* Some whalemen liked the fresh meat from the whale as a change from shipboard diet and they frequently used the best grade of the tried-out oil in which to fry fish caught in mid-ocean. The fresh oil was as acceptable as lard, and of course used to fry the whale steak and hamburgers. "In the long try watches of the night" wrote Melville in his *Moby Dick* "it is a common thing for the seamen to dip their ship-biscuit into the huge oil-pots and let them fry there for a while. Many a good supper have I made thus."

Men eat whales! and Jonah hails
The news with exultation!
It evens things and plainly brings
A due retaliation . . .
Jenkins' *Whales and Modern Whaling*

In the sixteenth century the meat of the whale was commonly marketed in Bayonne, Cibourre and Biarritz and salted blubber for table use was sold inland in the east of France. Beside the Basques, many other whaling peoples have appreciated the virtues of whale meat. It has long been eaten by the local inhabitants of certain areas of Norway, Greenland and Japan. In 1947 it was estimated that one-fourth of all the red meat in Japan was whale meat, most of it sold in cans. Long before it became an American dish for white men, whale meat was eaten by the American Indians. The *Illustrated World* of November, 1917, when the topic was much in discussion, pointed to a fast dwindling tribe of Indians on the Pacific Coast called Makahs, the only red people left who hunted the whale. The Makahs living at Neah Bay on the coast of Washington State used canoes, hand lances and spears. As many as twenty canoes, each carrying from four to ten huntsmen, joined in the chase. A whale became the property of the tribe and furnished the food supply for the entire village for months.

In appearance, the meat is of a closer texture than beef and its color is redder. In flavor, it can hardly be distinguished from prime beef. It lends itself to steak, roast, stew and makes admirable corned beef. All whales produce good meat, but the humpback whale, whose oil is poor, is said to have meat which is the most highly esteemed of all. An average whale supplies around 15 tons of choice tenderloin 15 feet thick. The great Rorquals of the Antarctic have a tenderloin 25 feet thick, which is wider than a city lot. Each whale yields some 40 tons or 80,000 pounds of edible meat. This is equivalent to the beef produced by 133 head of cattle. According to the American Meat Institute, a 1,000 pound steer produces 600 pounds of beef which includes by-products; 540 pounds become retail cuts. Little wonder that 1917 writers mused upon the whale's possibilities as a meat source. If the whale could only be domesticated, he could replace herds of cattle which require three years of maintenance, grazing and labor costs before they are ready for the stockyards. But a whale has never been captured alive. And even if he could be, he requires a ton of food a day to sustain him.

When beef prices became prohibitive during the meat shortage of World War I, the United States government advocated whale meat as a substitute. First placed on the market on the West Coast in April and May of 1917 at 10¢ a pound, it found a ready market. During 1917, 1,000 whales were taken in North Pacific waters. The catch represented 1,600,000 pounds of meat. In San Francisco, particularly, there was brisk demand for the whale beef. In New York, no less a personage than Dr. Henry Fairfield Osborne, noted naturalist-author and President of the American Museum of Natural History, championed whale beef to end the meat famine. Sometime in 1918, this gentleman gave a banquet to demonstrate the superiority of ocean venison. The menu of the now-forgotten feast still looks quite appetizing. It begins with Hors D'Oeuvre-Whale, celery and olives and proceeds to Whale Pot Au Feu. The entree is Planked Whale Steak, supported by vegetables, salad and dessert. Soon whale meat could be found on the menus of first-class hotels. Dr. Osborne would like to have seen 100 million pounds marketed here annually. Actually, only 1,600,000 pounds were sold which equalled the yield of 25,000 head of cattle.

Introduced during the first world war, whale meat continued to be used in the United States for human consumption. On the West Coast, whalers operated between Point Areans, California, up to the Oregon line. The captured whales were towed to Fields Landing and butchered by a land crew of thirty men. In 1929 the California Whaling Company handled whale meat while the Fields Landing plant prepared the tongue, organs, bones and inferior meat grades for the use of dog food factories. Once wasted to secure a trivial 3% of oil, whale meat became the most valued product of the whale at $800 a ton, or 40¢ a pound. In Britain, it sold for the same price as best Canadian beef. In 1947, 40 million pounds of whale meat were marketed in England, at 40¢ a pound, or $16 millions.

Much whale meat is ruined by use of the explosive harpoon gun. The damaged meat becomes oily and fishy to the taste, often complained of by the British. Whale meat will come into its own, only when the harpoon gun is replaced by a more efficient killing method equivalent to the quick dispatch of beef cattle. Meanwhile hundreds of millions of dollars worth of nutritious meat is wasted annually, sufficient to feed Europe for over three months at the rations prevalent in 1943.

Business Week of July 10, 1943 reported that the first refrigerator car shipments of 60,000 pounds of whale meat were rolling out of San Francisco for Eastern markets. The meat was processed by Maritime Industries, Eureka, California. Two San Francisco wholesale houses, Western California Fish Co. and

Producers Trading Co. handled national distribution. Butchers received the meat in 20-pound chunks on which the O.P.A. was asked to establish a ceiling of 35¢ a pound. The Maritime Industries also dehydrated the whale meat for human consumption, processing 12,000 pounds of fresh meat a day. It hoped to bring its daily output up to 30,000 pounds and expected to ship 3,750,000 pounds from a catch of some 125 whales. Government reports indicate that actually Maritime Industries caught only 29 whales in 1943 and 5 the following year, suspending operations in 1945 and 1946. In 1948, however, it secured 67 whales.

CHAPTER

6

Salem Begins in Earnest

"Fortune may find a pot but your own industry must make it boil." William Wilberforce (1759-1833)

In New England, whaling reached a high point by the time of the Revolution, when non-intercourse with Great Britain, Ireland and English West India islands disrupted progress and the Massachusetts ports felt the impact as much as the others. According to Felt, the industry was in a flourishing condition as early as 1765 when the whale fishery from Boston and the neighboring ports amounted to "100 sail which have been successful this season in the Gulf of St. Lawrence and the Straits of Belle Isle, having taken upwards of 9,000 barrels of oil." Losses were again suffered in 1799 because of American differences with France and again by reason of the War with England from 1812 to 1815.

Just as America began to whale on a large scale, a series of disasters befell the British whale fishery. The year 1830 saw nineteen British whalers destroyed in Arctic ice. After so great a misfortune, the industry fell on bad times. The shock to many investors was too great to be absorbed, and the British whaling industry entered its decline. By 1849 only 14 British whalers remained at work, instead of the 61 sent out from the port of Hull in 1821. Whaling honors so long divided between Britain and America were now to belong to the Americans who were to prosecute the whale fishery with vigor.

New England's golden years of whaling are placed between 1825 and 1860. New Bedford, the leading American whaling port, became the whaling metropolis of the world. The secret of New Bedford's phenomenal success was a persistent specialization in whaling. Other seaports of New England, Professor Morison points out, "hugged the delusion that foreign trade would return (after the War of 1812). New Bedford hugged her oil casks. Her Quaker shipowners who had made fortunes by neutral trading before 1812, perceived that the palmy days of the carrying trade were past, refitted their merchantmen as whalers, and went out after oil with a spirit and perseverance that made their town within six years the first whaling port of North America." (*Maritime History of Massachusetts,* p. 315.)

Fairhaven, which shares New Bedford's harbor, became by 1831 the third whaling center, while Nantucket was the second. Later Fairhaven was surpassed by New London. Edgartown on Martha's Vineyard had a fleet of 10 to 20 whalers in the forties and fifties. Provincetown, Cape Cod, at one time had as many as 30. The little seaports on Buzzard's Bay one by one entered the whaling game—Dartmouth, Mattapoisett, Marion, Wareham, Westport, Wood's Hole and Rochester. Blubber hunting touched almost every seaport of Massachusetts and Long Island Sound.

The citizens of Salem were kept informed of the merits of the industry by their newspapers. An illuminating article on the subject appeared on November 28, 1831 in the *Essex Register,* which gave the standing of the United States in the whale fishery in the current year. Over 300 ships, it announced, which averaged 340 tons each and requiring a total of 6,500 men to navigate them were now involved in whale catching. Many others were given employment in providing 6,000 tons of iron hoop for oil casks, 18,000 bolts of sail cloth, 36,000 barrels of flour, 30,000 barrels of beef and pork, 2,100 tons of cordage and whale line, as well as iron for harpoons, lances, spades, blubber hooks, etc. 1831 produced 112,000 barrels of whale-oil valued at 30¢ a gallon and 120,000 pounds of whalebone valued at 22¢ a pound. Whale-oil and whalebone were generally exported to Europe while sperm oil was used at home. A surprising amount of sperm oil found its way into candles. The report reveals the figure as 2,500,000 pounds of spermaceti candles produced, which gave employment to about eighty factories for refining oil and manufacturing the candles.

Salem merchants naturally hoped to participate in so lucrative a field. A few didn't mind adding whaling to their other mercantile interests. Strange as it may seem, Salem was to get its start in extensive whaling by reason of misfortune. In 1830 Salem suffered a business recession brought about by the opening of the Erie Canal and the introduction of the railroads. Both canal and railroads detracted transportation from Salem's harbor. That the town fathers were alarmed by these developments appears evident from articles in the Salem newspapers in November and December of 1830. Several communications to the editors had proposed the whale fishery as a means of recovering Salem from its doldrums. Apparently these appeals did not go entirely unheeded. Trade with India, Africa and China was beginning to wane and some shipowners began to outfit their vessels for whaling cruises and to invest quite heavily in the whaling industry.

In 1831 Salem registered the first whaler since the Brig *Nancy* sailed ten years earlier on a whaling-sealing voyage and returned with 1,800 seal furs and 100 barrels of whale-oil. On March 13th, 1831 Captain Hoit sailed for the South Atlantic whaling grounds in the Bark *Izette* built at Newmarket, New Hampshire. After the absence of one year and 11 days, the *Izette* returned with 100 barrels of sperm and 1,500 barrels of whale-oil. Salem proceeded cautiously. The next year it sent out 4 whalers, the Ships *Bengal, Catherine* and the *Izette,* and the Bark *Pallas.* The first two mentioned went to the Pacific whaling grounds as did the *Pallas.* The *Izette* retraced her paths to the South Atlantic where she had hunted the previous year.

By this time systematic efforts were under way by the Commonwealth of Massachusetts to place whaling on as firm a basis as fishing. This was to be achieved by paying bounties. To such an extent was the industry encouraged by subsidies that longer voyages were undertaken. Two-year voyages became standard practice and the far-off South Pacific the favorite hunting grounds of New England whalers. It took months for a sailing vessel to get into the Pacific. Before the short cut of the Panama Canal, the only route was the long one around Cape Horn. More often than not, it took three and four years to complete a whaling voyage to these distant waters.

In 1832 there were 139 whalers sent out by American ports. New Bedford was in the lead with 50; Nantucket had 25; New London, 12, Sag Harbor, 14, and the others less. In 1833, the port of Salem registered 8 whalers. According to Starbuck, the Brig *Charles Doggett* sailed for the Pacific on January 11th under Capt. Goodwin. The Ship *Catherine* left on a whaling voyage and the Ship *Clay* sailed on June 19th under Captain Church for the South Atlantic. The Bark *Emerald* under Capt. Eagleston sailed for the Pacific on November 23rd; the Bark *Eliza* for the same destination on May 26th. On May 15th the Ship *James Maury,* which had been bought from Boston for the purpose, sailed for the Pacific with Capt. Bigelow in command. The Bark *Reaper* sailed on September 7th for the Indian Ocean under Capt. J. T. Worth and the *Samuel Wright,* a full rigged ship, sailed this year under Capt. Pitman.

In 1834 Salem had several whalers absent on long whaling voyages and in this year John B. Osgood dispatched the Ship *Izette* under Capt. Sistare to the South Atlantic. After an absence of only a year, this vessel returned with a good cargo of 1,400 barrels of whale-oil. The active *Izette* which had sailed for whales in 1831, 1832, and 1834, made another voyage shortly after unloading her 1,400 barrels of whale-oil. Besides the *Izette,* six other whalers were registered out of Salem during the year of 1835. John B. Osgood was also responsible for the Ship *Bengal* which left for the South Atlantic in July for a two-year voyage; for the Bark *Reaper* to the South Atlantic and Indian Ocean which sailed in August and for the Ship *Lydia* which sailed in September for the South Atlantic whaling grounds.

The *Bengal, Reaper* and *Lydia* accounted for 2,740 barrels of sperm oil and 1,900 barrels of whale-oil. The value of the cargoes of these three whalers alone was some $88,560. Additionally, during 1835 the Bark *Cavalier* sailed under Capt. Russell for James King to the South Atlantic and sold 980 barrels of whale-oil at Rio Janeiro, returning with 75 barrels of sperm. The Bark *Richard* departed for the same location under Capt. Dewing for Joseph Hodges on October 12th, and the Bark *Palestine* under Capt. Cartwright sailed for Nathaniel Weston around Cape Horn to the Pacific on November 8th for a four-year voyage. All of these whalers made successful and safe voyages.

At this time the American whaling industry was in its stride with nearly thirty ports sending out whalers. More triumphs were to come as the United States achieved a record unequalled by any other nation. In the 1833-34 period, Britain still had 104 ships at sea in spite of the set-back suffered by her Arctic disasters of 1830. As for Salem, Massachusetts, there was every hope that it would take its place in the game with the other whaling ports. The upward trend was watched with satisfaction and it was with some local pride that an announcement appeared in the *Salem Gazette* on August 2, 1836:

> "The whaling business of this port has greatly increased since it was introduced till it has achieved a magnitude of which all of our citizens are not aware. We have now 15 vessels with an aggregate of more than 3,500 tons and involving a capital of $350,000. Hitherto the business has answered the expectation of those who are engaged in it and the success has been such as to induce further investments which are set up in contemplation. All of our whalers are now absent, two of the largest, the *Samuel Wright* and *James Maury* being expected home in the autumn. Two of

them are the individual property of Mr. Phillips, who we understand will soon add two more. The remainder are owned by companies."

In 1835 the Nantucket fleet went mainly to the Pacific and after 1840, it went almost exclusively there. Salem's whalers had been taking the long trek around Cape Horn since 1832 when the *Bengal, Catherine* and *Pallas* led the way. Most of Salem's whaling voyages since that time were to the Pacific Ocean. By 1850 the New Bedford fleet was to be found almost exclusively in these waters, now a prolific breeding ground.

Three other ports in Essex County were also active with whalers as Salem strove to get a firm footing in this industry. The fishing center of Gloucester sent out the Ship *Mount Wollaston* on two whaling voyages from 1833 to 1834, after getting a start in the late eighteenth century with the *Polly, Sea Horse* and *Two Friends.* Nearby Lynn, whose whaling doings are lost in obscurity, is known to have sent out its Ship *Atlas* on four whaling voyages between 1831 and 1835, its Ship *Clay* on three voyages from 1833 to 1836. Its Ship *Commander Preble* began her career of many voyages in the year 1834 and the Ship *Louisa* began hers in 1831 and the Ship *Ninus* in 1835. Lynn, however, was never active with more than seven vessels although they were all full-rigged ships and during the entire whaling period chalked up 33 voyages. Newburyport, also a whaling unknown, sent out its Ships *Adeline* and *Newburyport* both of which made a single voyage during the year 1834. This port's *Merrimac,* a ship, achieved four voyages from 1833 to 1842. During the whole whaling period Newburyport is credited with 13 voyages. Beverly, more active than Lynn, did not begin its whaling adventures until the year of the California Gold Rush.

CHAPTER

7

Whaling Merchants, Officers and Crews

My ventures are not in one bottom trusted, nor to one place;
Nor is my whole estate upon the fortune of this present year.
Shakespeare's *Merchant of Venice.*

Many well-known Salem shipping men appear as owners and investors in Salem's whaling adventures. Ownership was usually joint. Seldom were the ships owned outright by a single merchant although this sometimes was the case. Registration at the Custom House could be in the name of one of the owners, agent or business manager. In New England, it was usual for ownership in vessels to be in the form of shares in each individual whaler with some whaling firm designated as agent or manager. Only in the Hudson River whaling ports was it customary to have whaling *companies.*

Alexander Starbuck lists a single individual as "owner or agent." To determine who really owned any Salem whaler in question, one must resort to the *Salem Ship Register* which identifies the owners, gives the name of the master or captain and the date of registration, plus any interesting data available such as when the vessel was built, altered or purchased. Typical Salem names thus come to light rather unexpectedly related to the whaling industry.

Ownerships of some of the whalers chosen at random:

Ship *Elizabeth,* 397 tons. Built in Boston, 1827. Registered in Salem November 1, 1836 in the names of Stephen C. Phillips and James W. Cheever, owners; Isaac G. Hedge, Master. From 1836 to 1848 the *Elizabeth* was used as a whaler. In 1849 she was pressed into service for the California Gold Rush and was sold on the coast.

Ship *Emerald,* 271 tons. Built in Salem, 1823 and first registered in the names of John Forrester and Thomas H. Forrester, owners and James B. Briggs, Master. Her first registration as a whaler was on November 20, 1833 with Stephen C. Phillips as owner; John H. Eagleston, Master. Registered July 9, 1836 with James W. Cheever and Stephen C. Phillips as owners; Elisha Dexter, Master. Registered May 2, 1840 with Stephen C. Phillips, James W. Cheever and Robert Brown as owners and Robert Brown, Master. Registered September 2, 1843 with Stephen C. Phillips and James W. Cheever, owners; Edward F. Lakeman, Master. This whaler was shipwrecked on the coast of Madagascar in 1845.

Ship *Sapphire,* 365 tons. Built at Medford in 1825 and first registered in Salem in November, 1832, in the name of Stephen C. Phillips as owner and Samuel Grove, Master. Used as a whaler from 1836 to March 8, 1842, the date on which she foundered in the waters of the West Indies.

Schooner *Mac,* 80 tons. Built in Robbinston, Maine, 1831. Registered October 15, 1835 in the names of James Perkins, Samuel Simonds, John Howard, Jr., Stephen Roundy, owners; Stephen Roundy, Master. Registered as a whaler April 21, 1836 in the names of John B. Osgood, and "sundry other persons, owners;" James W. Winslow, Master. Registered in 1837 in the name of John B. Osgood; Captain Emmons.

Schooner *Franklin,* 89 tons. Built at Bucksport, Maine, 1823. Registered in Salem, April 6, 1836 in the names of James King and sundry other persons, owners; Alexander Newcomb, Master. Sold to Prospect, Maine in July of 1840.

Brig *Malay,* 268 tons. Built in Salem, 1818 and registered for the first time on August 24, 1818 in the names of Nathaniel Silsbee, William Silsbee, Zachariah F. Silsbee, Dudley L. Pickman, Robert Stone, owners; Moses Endicott, Master. Altered to a bark June 21, 1834. Registered as a whaler May 10, 1837 in the names of Stephen C. Phillips, Zachariah F. Silsbee, James W. Cheever, Edwin Barnard, Nantucket Island, owners; Edwin Barnard, Master. Registered March 21, 1842 in the names of James W. Cheever, Stephen C. Phillips, Zachariah F. Silsbee, owners; E. F. Lakeman, Master. Lost at sea in 1842.

Bark *Cavalier,* 294 tons. Built at Newmarket, New Hampshire in 1827. Registered at Salem for the first time on July 16, 1833 with Thomas P. Bancroft, owner, and Sylvester P. Fogg, Master. Registered as a whaler October 21, 1835 in the names of 32 persons: James King, Thomas P. Bancroft, William Stickney, Thorpe Fisher, Oliver Hubbard, William W. Palfray, John F. Allen, Isaac Cushing, Joel Bowker, George W. Jenks, Albert G. Brown, F. B. Crowinshield, Boston, Daniel Stoddard, Nathaniel J. Lord, Aaron Perkins, Morse & Crocker, Alexander Donaldson, John Archer, Samuel Simonds, Samuel Grant, Ezra Osborn, William Ball, Thomas Downing, Benjamin Brown, Theophilus Sanborn, George West, Richard Savory, Charles F. Putnam, William and Thomas Rea, Samuel Leach, Asa Sawyer, owners; Timothy Russell, Master. Sold to Stonington owners in April, 1845.

Ship *Bengal,* 304 tons. Built in Salem, 1816. Registered for the first time October 19, 1816 with Pickering Dodge as owner and Thomas Dennis as master. Registered as a whaler in 1832 with more owners and again on July 24, 1840 with John B. Osgood, owner; Benjamin Jackson, Master. Sold to New London 1844.

Ship *Lydia,* 292 tons. Built in Portsmouth, New Hampshire, 1821. Registered as a whaler September 21, 1835 with John B. Osgood as owner and William Ramsdell as master. July, 1837, same. After this voyage the *Lydia* was taken over by Nathaniel L. Rogers et al. Sold to Boston February, 1842.

Bark *Richard,* 252 tons. Built in Salem in 1826. First registered March 3, 1826 in the names of Joseph Hodges, John

Hodges, owners; John Hodges, Master. Registered October 13, 1835 in the name of John Hodges, Agent for the American Whaling Company, owner; Joseph Dewing, Master. Registered as a whaler August 1837 in the name of Joseph Hodges; Captain Dewing.

The Bark *Palestine,* 248 tons, had an unusual ownership. In 1835 it was procured by an association of some fifty-eight citizens interested in making Salem a whaling port.

As noted in some of the above registrations, the captain of the whaler was frequently one of the owners. No arrangement could give a man a greater incentive to take out a ship on a long and hazardous voyage than actually to own the whole or part of the vessel. He knew he would share in direct ratio to the success of the venture. The system of sharing permeated every whaling cruise from captain to the humblest member of the crew.

It is interesting that the Bark *Malay* was originally used in the pepper trade with Sumatra. Her master of 1818, Captain Moses Endicott, also served as master of the Brig *Persia* which imported 160,000 pounds of pepper consigned to Dudley L. Pickman in July, 1824. In 1836 the *Malay,* Captain Silsbee, returned to Salem from a pepper voyage to Sumatra. She was then outfitted as a whaler and her next voyage was to the South Atlantic and Indian Ocean. Her owner was the prominent shipowner and merchant Stephen C. Phillips.

A son of Captain Stephen Phillips, Mr. Phillips was the father of Stephen H. Phillips, the Attorney General for Hawaii, and grandfather of both Stephen W. Phillips, the present President of the Essex Institute in Salem, and of his brother, James Duncan Phillips, the Salem historian. Stephen C. Phillips became Salem's second mayor when he succeeded Leverett Saltonstall in 1838. At this time, he was represented in Salem's whaling industry by the two Ships *Elizabeth* and *Sapphire* and the Barks *Emerald, Eliza, Henry* and *Malay.*

Analysis of sailings of all whalers from the port of Salem

between 1834 and 1868 reveals that there were 22 whalers registered by John B. Osgood, 14 by Stephen C. Phillips, 6 by John C. Osgood, the son of John B. Osgood, 5 by Benjamin Webb, 4 by James King, 3 by James W. Cheever, 2 each by Joseph Hodges, Nathaniel Weston and John B. Pierce and 1 by Timothy Bryant, Jr.

Many Salem captains who had completed perilous voyages in uncharted seas, transferred their talents for adventure to Salem whalers. In some cases, the whaling merchants engaged experienced whaling masters and officers from other whaling ports. The Nantucket captains, for instance, have become familiar names in Massachusetts marine history. There were Worths, Ramsdells, Barnards, Husseys, Coffins, Chadwicks, Colemans, Browns and Starbucks. Some of them appear as masters of Salem vessels. It would take a volume larger than this to clear up the historical mist which surrounds the personalities of the men themselves who engaged in Salem whaling, the families from which they came and the ports for which they sailed during their seafaring years.

The owners at home were often involved in relative industries such as rope making, iron works, cordage, shooks for barrel staves, manufacturing of sail cloth, provisioning of the ships, etc. The sailing of a whaler represented the joint efforts of hundreds of men. A great deal of expense was involved before the high point of a whale hunt was reached with the harpooner standing with arm raised, nerves taut, ready to toss his weapon.

Much depended upon the whaling captain. He required the usual navigational ability, judgment, perseverance and seamanship. Additionally, he needed experience and that indescribable special "fisherman-sense" which enabled him to know where the whales were. It took years of roaming the seas and encountering whales to acquire the intangible equipment of a successful whaling captain, and still most of these mariners were seasoned salts before they had reached the age of thirty. First and foremost the master of a whaler needed the knowledge of navigation to bring the vessel safely through vast expanses of ocean to the whaling grounds, then he needed the talents necessary to further the actual work of the look-outs and harpooners in securing the objects of the search, and then to oversee the great task of reducing the whale to oil while at sea.

The importance of the officers cannot be overrated. If the whaling crew lacked confidence in the "old man" or if the harpooner lacked experienced skill, or muffed his chances at fastening to the whale once too often after the backbreaking chase, then the crew would lose all heart in the cruise, would become bored, troublesome, impudent, or even desert ship at the first opportunity. Nothing was so disheartening to the captain and officers than a fruitless search of months or weeks with no chances for catching whales. The idle men lived on hope from day to day and finally became disgruntled and often defiant, blaming the failure of the whales to appear on the captain and officers. The old logs are peppered with instances of this kind and even of insubordination. John Nolan, for one, became antisocial during the 1838 voyage of the Salem Ship *Elizabeth*. In the words of the logkeeper, this is what happened: "*Sept. 6, 1838* on board the Whaler *Elizabeth:* John Nolan being called to go aft refused and said he would not go by the captain's orders. He would not go until compelled by taking hold of him to bring him aft. Nolan laid his hands upon the captain and refused to go but being overpowered, he went and was put in irons, denying all duty and giving much insolence and he swore he would not hold his tongue after being told to do so by the captain, and began singing all sorts of lascivious songs and bidding defiance and said that he had only come on a voyage of pleasure and many abusive things."

"*Sept. 7.* John Nolan broke off his irons and attempted to go forward but was intercepted by the third mate, although Nolan threatened to put a knife into the man who laid his hands on him. He also told me he was sorry he had not lain the captain a corpse while he was bringing him aft to put him in irons.

At 7 P M the captain called him and he acknowledged his faults, and said he was ready to go on duty."

I can find no entries in the Salem logs in which a captain or officer was killed by a member of the crew, although Salem skippers and mates perished during long voyages from hardships and disease. Sanguinary tales in whaling annals refer to such exceptional voyages as the Whalers *Glove* and *Junior* of New Bedford. In both cases most of the officers were murdered in cold blood by the whaling mutineers.

It is not always possible to get the captain's viewpoint during a voyage as many skippers left the keeping of the ship's log to the mate. However, the log of the 1838 voyage of the *Reaper* from Salem was kept by Captain Neal and gives some rare insights into a whaling master's problems. After having his waist boat stove into "toothpicks by a whale," losing 22 irons and 1 lance in four whales which got away, a whale was finally captured and brought alongside. They got the whale "stewed out" but whether the oil "will sell is more than I know" lamented Captain Neal "but I do know that it looks more like tar than oil. This is probably the finishing of our Right whaling and I must in all sincerity confess that I don't care to go a Right whaling again. I shall now try Fort Dauphin. If I can take some Sperm there, it will be a God-send. If I am doomed to bad luck to the end of the voyage, give me strength to bear it. My upper lip is dreadful limber at present and 200 barrels of sperm oil would be an excellent stiffner for it." (Nov. 26, 1838.)

The absence of whales was disappointing, of course, but the resultant attitude of the crew was an even greater cause for anxiety. Little wonder that tempers on board a luckless ship now and then burst into flames of revolt. A situation approaching mutiny is described in the log entry for Christmas day of 1839. "Stormy with rain. This morning the whole of my crew came aft to know when I intended to go into port, saying that if I did not go in soon they would not work the ship. The weather is now bad and it is not a suitable time to make any port in this vicinity. If I am forced to go into Fort Dauphin both the crew and ship should be lost. It must be put to their charge to say the best thing . . . They are a turbulent set of men, entirely unfit for a whaleship and it is very unfortunate for a ship to have such men. I have heard of several ships being forced home by their crews. It has been my hard fate to have had as much trouble with this set of men as would suffice for half a dozen ships, and yet they can find no fault with the wage on board. This is more than I can say. I shall try to get into Fort Dauphin, Madagascar, as soon as the weather will permit."

The next day Captain Neal was comforted by meeting up with and speaking the Whaler *Martha* of Nantucket. The *Martha* reported the loss of four of her men by sickness at St. Mary's. At this time there were two men sick on board the *Reaper* and unfit to work. On December 26, 1839, the *Reaper* log revealed a situation which could have terminated the voyage. "I have been told" recorded the captain "that there is an agreement between some of my men not to report a whale if they see one. If this is a fact, and I have reason to believe it is from some circumstances which have transpired within two days, it would be better for all concerned that the ship should go home. It is worse than useless for me and a few others to fret and worry and try hard to get oil, when the bulk of my men are worse than bad. To meet with success in a whaling voyage, it is of the utmost importance that all on board should haul together. It is a business that needs the utmost exertion when one has oil to take care of, the whole mind and body must be given to it, if not, it had better be let alone."

The log tells of making port and remaining there for twelve days, departure having been delayed by lack of wind for the sails. The discontented crew demanded certain privileges of the captain which he granted with much reluctance. The following night they repeated their demands and tried to force the Captain to fulfill them. "I have told them" says the log "it cannot be done. I have tried to reason the case with them but it is of no use. They have combined together tonight that they will do no more duty. Mr. Hersey heard the agreement. It was this: that when they were called in the morning they would say that they were sick. The mates and boatsteerers stand the watch tonight. The wind is strong."

At the time the men were denied liberty, the *Reaper* was anchored off-shore in the harbor of Fort Dauphin. Well, if they could not get to their grog shops and brothels, they were determined that these be brought to them. It was this request Captain Neal was compelled to grant to his men. He noted in his journal that he "slept on shore and let the girls come on board for the purpose of keeping the crew quiet." He admitted that "this was hard, but I am forced to do it." When the captain denied the men a repeat visit, they schemed to deny *him* any work. At four o'clock in the morning, a light wind arose and the *Reaper* finally pulled out to sea and away from temptation. Eight of the men remained rebellious. They did no work and had to be watched.

Most crews were kept in a satisfied frame of mind by the intermittent catching of whales, and an occasional liberty when the whaler made a port. If a captain could, he would try to avoid the unpleasant aftermath of officers and men imbibing too freely and overstaying their liberty sometimes as much as 48 hours. Captain Neal notes with some bitterness in his log for March 7, 1839. "There is a schooner here at Fort Dauphin which has been detained three days by the wind. She undertook to go this morning, but could not. If I can get this Barque out, it will be after this when I come here again in a whaleship. More especially, in fact, a whaleship ought, if possible, to avoid all ports in which rum can be obtained, for it is a fact not to be disputed that the generality of seamen will sell the last rag of clothing to buy rum without the least thought of the morrow.

You may have temperance societies and seamen to fill the list of names, then publish it in your report, but it is but empty brass and a tinkling sound."

The fortunate capture of whales put new heart in a cruise and spurred the men's hopes of a bonanza voyage. There was one thing which often soured the men. If the long search ended in losing the prize by a clumsy or nervous harpooner, then the captain was in a difficult situation. The loss became conversation piece for the men for days on end. If fate did not send another whale in good time, the spirit of the cruise was ruined. The men became demoralized, ugly or glum. Sometimes a crew became so disheartened that when the test of strength and daring arrived, it was impossible for them to rise to the occasion.

On the other hand, nothing succeeds like success. Clever harpooning, bravery of captain and officers in action, engendered confidence. The men became more sure of monetary rewards. The skillful capture of a whale or two boosted the ship's score. The team was satisfied that the harpooner knew his business and gave him their support regardless of their own personal safety. Loyalty to worthy leadership was often so strong that the loss of a limb or a broken rib did not change the victim's mind. He consoled himself that the sacrifice was worth it. It cannot be denied that there was a personal element in whaling which was very important.

CHAPTER

8

Brig, Schooner, Bark and Ship

Comparable only to the prairie schooner, the whaleship will always remain an American epic symbol.
(Franklin D. Roosevelt's Preface to Clifford Ashley's *Whaleships of New Bedford.*)

It was the rigging of the vessel of the old sailing days which designated her classification. The mention of bark, brig, schooner etc. enabled a man to envision her appearance. The nine principal types of American sailing vessels were each different from the other by reason of its rig from the simple sloop with its single mast and fore-and-aft rig to the full-rigged ship of three masts. The rigs mentioned in connection with Salem's whalers are the brig, schooner, bark and ship.

Of the thirty-eight vessels involved in Salem's whaling industry, fifteen were full ships, thirteen were barks, seven were brigs and three were schooners. The *brig* is a vessel with two masts, fore and main, both of which are square-rigged. On the mainmast there is a standing gaff to which is rigged a small fore-and-aft sail. In all three classes of brigs the foremast is made in three spars and square-rigged, but the mainmast is differently rigged in each type.

The *schooner* is a vessel of two or more masts, fore-and-aft rigged with the fore- and mainsails suspended from gaffs and laced to booms on the foot of the sails. The earliest type had only two masts which were later increased. The three-masted schooner was the most popular type in the nineteenth century. The topsail schooner had two masts, the mainmast of which has a fore-and-aft mainsail and gaff topsail. Both masts are made in two spars, but the lower foremast is a little shorter than the corresponding spar of the mainmast and the topmast is a little longer.

The *bark* is a three-masted vessel with foremast and mainmast square-rigged and the mizzenmast (third mast) fore-and-aft rigged. The mizzenmast carries no yards; there is a hoist-and-lower fore-and-aft sail and a gaff topsail.

The term *ship* is properly used only in designation of the full-rigged ship, a large square-rigged vessel which carries three masts, each of these being composed of a lower-mast, topmast and top-gallant-mast and is provided with yards and carries a full complement of square sails. After 1850 topsails were generally divided into upper and lower for convenience in handling.

Many of Salem's whalers were adapted from commercial carriers. Alterations also occurred from one rig to another as in the case of the Schooner *Para* altered to a brig. The Brig *Malay* first registered in 1818 advanced to a bark in 1834. The alteration did not always take a forward movement toward a full-rigged ship. For instance there was the Ship *Eliza* which was altered to a bark in 1838. The Ship *Emerald* was also altered to a bark in 1836. The changing of the rig from full ship to a bark was in the interest of economy. It took 35 to 38 men to operate a ship. The new rig of a bark could be handled with less than 30 men. If the ship were not too large in the first place, these figures would vary in the right ratio.

Old merchantmen of Salem were early in demand by whaling interests of other seaports, who readily converted them into whalers. There was the tale of the Salem carrier which became a New Bedford whaler told by William H. Nichols, who kept the log of the *Emerald*. "The *Hercules*" wrote Mr. Nichols in a memorandum which is now pasted in the log of the *Emerald*'s 1838 voyage "was a very fine looking old ship with her bright waist. She was a fast sailer and formerly belonged to Salem, being an old East-India man. On her voyage to Batavia, Java, previous to 1816, her carpenter John Nichols died there, an uncle of the writer. In those days many old Salem ships were sold to New Bedford parties for whalers. The Salem Ship *Hope* was one and on her first whaling voyage, our Mate Downes was on her, and I heard him say that every time they pumped ship, they would pump up lots of pepper, she having been in the Sumatra trade while owned in Salem." The Brig *Roscoe* of Salem was another vessel known to have been sold to New Bedford for a whaler in 1832.

Some of the vessels used as whalers were bought by Salem merchants from other ports. The Ship *Elizabeth* was built in Boston in 1836, the year she was registered as a whaler for the first time in the name of Stephen C. Phillips. The Ship *Emerald* likewise registered in the same year, appears to be a native of Salem of 1823. Mr. Phillips' *Sapphire* hailed from Medford, the Schooner *Mac* from Robbinston, Maine, the Schooner *Franklin* from Bucksport, etc.

The outstanding characteristic of a whaler was her strength. The thickness of the planking, timbers and lining was planned for endurance. Her second characteristic was space for storing hundreds upon hundreds of oil barrels. The great cavities needed for storage precluded the classic lines and beauty of the merchant marine. When she sailed the seas, her critics said she went *lumbering along* and I suppose she did. The whaler was typical of the square-riggers, bluff of bow, roomy in the bilges for the products of the whale, square in the transom and sparred for endurance rather than speed. The Ship *Elizabeth* built at Boston had two decks and three masts, was 113′ 2″ long, 28′ broad, 14′ in depth and displaced 397-43/95 tons. Her stern was square. There were no galleries. Nevertheless, she did have a touch of the sublime. On the bow of the *Elizabeth* there was a figurehead. The carved and painted figureheads of the old sailing ships ranged from life-sized ladies of high fashion to unclothed mermaids with tails and breast tips of gold. There were golden eagles and sea gods and goddesses.

It was easy to recognize a whaler by the peculiar construction of the *after-house* called by sailors the *round-house* which contained the cook's galleys, locker rooms and the cabin gangway and a shelter of sorts for the helmsman. Peculiar to whalers were the wooden *davits* or cranes used for suspending, hoisting or lowering the whaleboats. There was the crow's nest placed high up on the main topmast or t'gallant-mast, made of canvas or light wood from which the careful look-out was kept for whales. No other ship carried fiery furnaces on deck which gave off thick black smoke. Then, too, the encircling white border proclaimed the American whaler.

If the old sailing whaler is a thing of the past, so is the complement of her being—the long, slender, handsome whaleboat. What the whaler lacked in grace was more than made up for by the whaleboat. With the innovation of the harpoon gun, the steam "catcher" of today's hunt, crews are no longer lowered into the light rowboats for the long chase and capture. For many years, there has been no exposure of whalemen to the risks and grave dangers which were part and parcel of the old whaling years. It would be wrong to omit mention of the whaleboats, the stage upon which all the roles of heroism, skill and even death were enacted.

Regulation size for a whaleboat was 28′ long, 6′ wide at the center thwart, and sharp at both ends for speed in either direction. She was built with oak ribs and half-inch cedar planks, was big and light. The design can be traced right back to those early Viking craft which were from 25′ to 28′ long and about 5½′ in the beam. Buoyant and graceful in her movements, the swift spear leaped from billow to billow, appearing to dance over the sea, rather than to plough its bosom with her keel. Each boat could carry twelve persons so that in the event a companion boat was *stoved* or pierced by the whale, she could go to the rescue and then continue the chase. Regulation crew numbered six men and up to the time the whale was fastened, every man except the boatheader pulled a long oar 12 or 14 feet while the boatheader steered with his 28′ long steering oar. In each boat were two tubs of harpoon line of the best possible cordage made of tarred hemp. The whale could use up a quarter of a mile of line before the second tub's cordage needed to be called on.

In the back of the whaling logs, the mate kept his scores of whales which were caught, those that got away and how many whaleboats were destroyed in the attack. The whale had a code of contest all his own. His chief defense was to immediately take the offensive: to shatter the rowboat and to dump his enemies in the briny deep. The routine which is described in Melville's *Moby Dick* was the basic pattern of whale pursuit.

When a whale was seen it was the look-out's first duty to "sing out" and then on the word to "lower away," Captain Ahab would always ask: "What tune do you pull to?" and the men would answer with a shout. "A dead whale or a stove boat!!!" A lost boat was a calamity duly recorded. A lost whale was indicated in the margin of the log book by a drawing of a whale's tail. The lost human being killed in conflict or drowned is indicated and as like as not mentioned no more except for the auctioning off of his possessions to the crew, the money to go to his widow eventually—or not.

Beginning with the ill-fated Salem Ship *Britannia,* which was wrecked before she could clear the harbor, whaling voyages out of Salem commence with the year 1818. Here is included a record of the voyages and the types of vessels which were engaged in them. It would take volumes to encompass all of the daring, courage, triumphs and heartaches bound up in these simple listings touching a world of wooden ships of sail against the sea and of whalemen with their hand harpoons against their quarry. One of America's lost industries, whaling in the days of sail remains part of the national heritage:

1818 *Britannia,* Ship, sailed Dec. 5, wrecked on Picard's rocks, Manchester, going out. No further report.

1820 *General Knox,* Ship, Capt. Orne, to Falklands, arrived home June 6, 1821. On a sealing voyage principally, brought home 5,000 seal skins and 600 barrels of whale-oil.

Polly, Brig, left in August. No further report.

1821 *Nancy,* Brig, Capt. Upton, managing owner, S. White, to Falkland and N. S. Shetlands. Arrived home May 27, 1822 with 100 barrels of whale-oil and 1,800 seal skins.

1831 *Izette,* Bark, Capt. Hoit, to South Atlantic. Sailed March 13, arrived home March 24, 1832, with 100 barrels of sperm and 1,500 whale-oil. Built at Newmarket, New Hampshire.

1832 *Bengal,* Ship, 304 tons, Capt. Russell, to Pacific Ocean. Sailed Mar. 24. Returned Feb. 23, 1835 with 1,200 barrels of sperm.

Catherine, Ship, no tonnage given, Capt. Paddock, to Pacific. Sailed Mar. 24.

Izette, Ship, 275 tons, Capt. Kempton, to South Atlantic. Sailed June 9.

Pallas, Bark, no tonnage, Capt. Archer, to Pacific. Sailed July 21.

1833 *Charles Doggett,* Brig. Capt. Goodwin, to Pacific. Sailed Jan. 11.

Catherine, Ship, burned off Oahu in 1834. (This fate according to Starbuck, befell the *Catherine* and not the *Charles Doggett* as sometimes alleged.)

Clay, Ship, Capt. Church, to South Atlantic. Sailed June 19.

Emerald, 271 tons, Bark, Capt. Egleston, owner or agent, John B. Pierce, to Pacific. Sailed Nov. 23.

Eliza, Bark, 262 tons, Owner or agent, James W. Cheever, to Pacific. Sailed May 26.

James Maury, Ship, 395 tons, Capt. Bigelow, Managing owner or agent, John B. Osgood, to Pacific whaling grounds. Sailed May 15. Bought from Boston 1833.

Reaper, Bark, 230 tons, Capt. J. T. Worth, to Indian Ocean. 900 bbls of whale-oil. September 7. Returned July 5, 1835.

Samuel Wright, Ship, 372 tons, Capt. Pitman, Managing owner or agent, returned home August 27, 1836 with 2,000 barrels of sperm.

1834 *Izette,* Ship, 275 tons, Capt. Sistare, Managing owner or agent, John B. Osgood, to South Atlantic. Sailed May 21. Arrived home April 21, 1835 with 1,400 barrels of whale-oil.

1835 *Bengal,* Ship, 304 tons, Capt. George Netcher, Owner or agent, John B. Osgood, sailed July 11 to South Atlantic, returned March 28, 1837 hailing 140 barrels of sperm and 1,600 barrels of whale-oil.

Cavalier, Bark, 295 tons, Capt. Russell, Managing Owner or Agent, James King, to South Atlantic, sailing October 25, returning May 22, 1837 with 75 barrels of sperm. Sold 980 barrels of whale-oil at Rio Janeiro.

Reaper, Bark, 230 tons, Capt. Jackson, Managing Owner or Agent, John B. Osgood, to South Atlantic and Indian Oceans, sailing Aug. 5, returning July 12, 1837 with 1,100 barrels of sperm oil.

Richard, Bark, 252 tons, Capt. Dewing, Managing owner or agent, Joseph Hodges, to South Atlantic, sailing Oct. 12, returned in 1837.

Lydia, Ship, 293 tons, Capt. Ramsdell, Managing Owner or agent, John B. Osgood, to South Atlantic, sailing Sept. 22, returning Nov. 5, 1837 with 1,500 barrels of sperm-oil and 300 barrels of whale-oil. Vessel bought from Portsmouth, 1835.

Izette, Ship, Capt. Sistare, South Atlantic, returning in Nov.

Palestine, Bark, 249 tons, Capt. Cartwright, Managing owner or agent, Nathaniel Weston, to Pacific Ocean, sailing Nov. 8, returned on April 10, 1839 with 1,600 barrels of sperm. Vessel bought from Boston in 1835.

1836 *Elizabeth,* Ship, 397 tons, Capt. Hedge, Managing owner or agent, Stephen C. Phillips, to Pacific. Sailed Nov. 5, returned May 1, 1840 with 2,400 barrels of sperm. Bought from Boston in 1836.

Emerald, Ship, 271 tons, Capt. Dexter, managing owner or agent, Stephen C. Phillips, to Pacific, sailed Nov. 5, 1836 and returned May 1, 1840 with 300 barrels of sperm and 1,450 barrels of whale-oil. Built in 1824.

Statesman, Bark, 258 tons, Capt. Coffin, managing owner or agent, Timothy Bryant, Jr. to Pacific, sailing Dec. 22, returned Sept. 23, 1838 with 2,100 barrels of sperm.

Samuel Wright, Ship, 372 tons, Capt. Coffin (relative?) managing owner or agent, John B. Osgood, to Pacific, sailing Nov. 24, returning March 1, 1839 with 300 barrels of sperm and 2,200 barrels of whale-oil. Built in 1831.

Sapphire, Ship, 366 tons, Capt. Mayhew, managing owner or agent, Stephen C. Phillips, to Pacific. Sailed June 19 and returned Sept. 9, 1839 with 1,000 barrels of sperm and 500 barrels of whale-oil. Built in 1820.

Mount Wollaston, Ship, 325 tons, Capt. Jewett, Managing owner or agent, John B. Osgood, to South Atlantic, sailing June 28, returning April 12, 1838 with 450 barrels of sperm and 1,250 barrels of whale-oil. Built in 1822.

Mac, Schooner, 80 tons, Capt. Winslow, managing owner or agent, John B. Osgood, to Atlantic, sailing April 21, returning April 23, 1837 "clean," cargo disposed of. Built in 1831.

Emeline, Brig, 98 tons, Capt. Lombard, managing owner or agent, John B. Pierce, to the Atlantic, sailing Mar. 28, returning on June 8, 1837 with 75 barrels of sperm and 20 barrels of whale-oil. Built in 1828.

Franklin, Schooner, 89 tons, Capt. Newcomb, owner or agent, James King, to Atlantic. Sailed April 6, returned May 9, 1837 with 40 barrels of sperm. Built 1828.

1837 *Bengal,* Ship, 304 tons, built 1816. Capt. Jackson, owner or agent, John B. Osgood, to South Atlantic and Indian Ocean. Nov. 6 to May 29, 1840. 1,800 barrels of Sperm.

Cavalier, Bark, 295 tons. Built 1827. Capt. Francis, owner or agent, James King, to South Atlantic and Indian Ocean. Sailed July 8 and returned March 24, 1839 with 180 barrels of sperm, 820 bbls. of whale-oil, 8,000 pounds of whalebone. Sold out 1839.

Derby, Bark, Capt. Radcliff, no owner or agent given, to Indian Ocean. Sailed July 15. Lost at Falkland Islands, April 15, 1838. Shipped oil home.

Emeline, Brig, 98 tons, Capt. Lombard, owner or agent John B. Pierce to Atlantic Ocean. Sailed Sept. 5, returned July 1, 1838, with 110 bbls. of sperm, 5 bbls. of whale-oil. Sold out 1838.

Franklin, Schooner, 89 tons, Capt. Tracy, owner or agent, James King, to Atlantic, sailing on May 27th and returning December 8, 1837, "clean." Sold out 1837.

Izette, Ship, 275 tons, Capt. Hall, owner or agent, John B. Osgood, to South Atlantic and Indian Ocean, sailing October 16, returning December 20, 1839, with 250 bbls. of sperm and 2,050 of whale-oil. Built 1825.

James Maury, Ship, 395 tons, Capt. Bigelow, owner or agent, John B. Osgood, to South Atlantic and Pacific Ocean, sailing July 5 and returning Dec. 11, 1840 with 400 barrels of sperm and 2,600 barrels of whale-oil. Built 1832.

Lydia, Ship, 293 tons, Capt. Ramsdell, owner or agent, John B. Osgood, to South Atlantic and Indian Ocean, Dec. 20, returning March 25, 1840 with 380 barrels of sperm and 1,450 of whale-oil. Built 1822.

Malay, Bark, 268 tons, Capt. Barnard, owner or agent, Stephen C. Phillips. To Indian Ocean, May 21, returning May 5, 1839 with 500 bbls. of sperm and 1,000 of whale-oil.

Mac, Schooner, 80 tons, Capt. Emmons, owner or agent, John B. Osgood, to Atlantic, sailing May 24 and returning Nov. 17, 1837, with 70 bbls. of sperm.

Reaper, Bark, 230 tons, Capt. Neal, owner or agent, John B. Osgood, to South Atlantic and Pacific Ocean, sailing August 21 and returning May 28, 1839 with 500 barrels of sperm and 100 barrels of whale-oil. Built 1825.

Richard, Bark, 252 tons, Capt. Dewing, owner or agent Joseph Hodges, to South Atlantic and Pacific, sailing April 19th, returning March 1, 1839.

Eagle, Brig, Capt. Williams, sailed as tender to *Richard.* *Richard*'s oil sold at Pernambuco. Tender *Eagle* lost in July or August off Montevideo in the merchant service.

1838 *Eliza,* Bark, 362 tons, Capt. Radcliff; owner or agent, James W. Cheever, to Indian Ocean, sailing Nov. 21 and returning May 9, 1841 with 275 barrels of sperm and 1,300 of whale-oil.

Emerald, Ship, 271 tons, Capt. Dexter; owner or agent, Stephen C. Phillips, to Indian Ocean, sailing May 24 and returning February 27, 1840 with 250 barrels of sperm and 1,750 of whale-oil. Sold 100 bbls. sperm at Hobart Town, Tasmania, Australia.

Mount Wollaston, Ship, 325 tons, Capt. Jewett; owner or agent, John B. Osgood, to Indian Ocean, sailing June 2 and returning April 17, 1840 with 600 barrels of sperm and 1,100 barrels of whale-oil.

Statesman, Bark, 258 tons, Capt. Coffin; owner or agent, John B. Osgood, to New Zealand, sailing October 26 and returning March 9, 1842 with 1,800 barrels of sperm.

1839 *Malay,* Bark, 268 tons, Capt. Barnard, owner or agent, Stephen C. Phillips, to Indian Ocean, sailing July 3 and returning January 20, 1842 with 1,300 barrels of sperm.

Palestine, Bark, 249 tons, Capt. Crimblish; owner or agent, Nathaniel Weston, to Indian Ocean, May 18, returning October 16, 1842 with 1,700 barrels of sperm. The Captain and First Mate, George Coffin, died in 1841. The *Palestine* was wrecked after this voyage. Built 1835.

Reaper, Bark, 230 tons, Capt. Neal; owner or agent, John B. Osgood, to Indian Ocean, sailing May 24 and returning Sept. 13, 1842 with 1,000 barrels of sperm.

Samuel Wright, Ship, 372 tons, owner or agent, John B. Osgood, sailing May 4, this whaler never returned. Lost at New Holland July 8, 1840.

Sapphire, Ship, 366 tons, Capt. Cartwright; owner or agent, Stephen C. Phillips, to Pacific. Sailed Nov. 28 and returned December 17, 1842 with 800 barrels of sperm and 1,200 barrels of whale-oil. Foundered at sea after this voyage bound to Mobile, West Indies 1843.

1840 *Bengal,* Ship, 304 tons, Capt. Jackson; owner or agent, John B. Osgood, to Indian Ocean, sailing July 24 and returning March 26, 1844 with 1,800 barrels of sperm oil. Sold to New London in 1844.

Emerald, Bark, 271 tons, Capt. Brown; owner or agent, Stephen C. Phillips, to Indian Ocean. Sailed May 2 and returned Feb. 26, 1843 with 400 barrels of sperm, and 1,100 barrels of whale-oil and 8,800 pounds of whalebone.

Izette, Ship, 275 tons, Captain Hall; owner or agent, John B. Osgood, to Indian Ocean, May 2 and returned on June 19, 1842 with 900 barrels of sperm and 1,100 of whale-oil.

Mount Wollaston, Ship, 325 tons, Capt. Rose; owner or agent, John B. Osgood, to Indian Ocean, June 24, returning June 11, 1843, with 400 barrels of sperm, 700 barrels of whale-oil and 5,600 pounds of whalebone. Sold 1,200 barrels of whale-oil at Rio Janeiro. Vessel sold to ——— 1843.

1841 *Eliza,* Bark, 262 tons, Capt. Chase; owner or agent, James W. Cheever, to Indian Ocean, sailing July 3. 200 barrels of sperm (probably sent home). Condemned at Tahiti, July 1843.

Elizabeth, Ship, 398 tons, Capt. Hedge; owner or agent Stephen C. Phillips, to the Pacific, sailing January 12 and returning Oct. 20, 1844 with 1,500 barrels of sperm.

Henry, Bark, 262 tons, Capt. Manchester, no owner or agent given, to Indian Ocean, sailing Oct. 14 and returning April 15, 1845 with 140 barrels of sperm, 300 barrels of oil and 2,400 pounds of whalebone.

James Maury, Ship, 395 tons, Capt. Benjamin R. Hussey; owner or agent, John B. Osgood, to Indian Ocean, sailing March 11 and returning February 19, 1845 with 1,400 barrels of sperm, 500 barrels of whale-oil and 3,600 pounds of whalebone. Sold to New Bedford 1845. Captain Hussey died June 15, 1844 and Charles F. Pinkham, First Mate, died September 1844.

1842 *Malay,* Bark, 268 tons, Capt. Lakeman; owner or agent, Stephen C. Phillips, to Indian Ocean, sailing March 26. Lost in Mozambique Channel, July 1842.

Statesman, Bark, 258 tons, Capt. Elisha Doane; owner or agent, John B. Osgood, to Pacific. Sailed June 12 and yielded 500 barrels of sperm oil. Condemned at Talcahuano, Nov. 1844.

1843 *Emerald,* Ship, 271 tons, owner or agent, Stephen C. Phillips, to Indian Ocean. Sailed Sept. 3. Voyage yielded 1,100 barrels of sperm oil. *Emerald* wrecked off Fort Dauphin, Madagascar, March 18, 1845. Oil saved.

1844 *Elizabeth,* Ship, 397 tons, Capt. Hall; owner or agent, Stephen C. Phillips, to Pacific, sailing Dec. 17 and returning May 15, 1848 with 620 barrels of sperm and 1,580 barrels of whale-oil. 14,000 pounds of whalebone.

1845 *Henry,* Bark, 262 tons, Capt. Lind; owner or agent, James W. Cheever, to Pacific, sailing June 12. Voyage yielded 300 barrels of sperm and 600 barrels of whale-oil. Vessel was wrecked at Marquesas Islands, got off and was taken to Tahiti and sold. 800 barrels of the oil was saved.

1851 *Margaretta,* Bark, 230 tons, Capt. Prior; owner or agent, Benjamin Webb, to Atlantic, sailing May 3rd and returning July 26, 1852 with 320 barrels of sperm and 37 of whale-oil.

1852 *Margaretta,* Bark, 230 tons, Capt. Holmes; owner or agent, Benjamin Webb, sailing Sept. 17 and returning Dec. 8, 1853. 180 barrels of sperm and three barrels of oil.

1854 *Messenger,* Bark, 216 tons, Capt. Holmes, owner or agent, Benjamin Webb, to Atlantic, sailing June 14 and returning August 30, 1856 with 230 barrels of sperm, 219 barrels of whale-oil and 1,200 pounds of whalebone. Withdrawn 1856.

1857 *Messenger,* Bark, 216 tons, Capt. Holmes, owner or agent, Benjamin Webb, to Atlantic, sailing April 29 and returning August 25, 1859, with 362 barrels of sperm and 114 barrels of oil.

1860 *Messenger,* Bark, 216 tons, owner or agent, Benjamin Webb, April 18, returning June 17, 1861 with 330 barrels of sperm and 16 barrels of whale-oil. Sold to Boston 1861. Became one of "stone fleet" No. 2 sunk in 1861.

1862 *Falcon,* Brig, 159 tons, Capt. Holmes; owner or agent John C. Osgood, to Atlantic, sailing May 19 and returning July 26, 1863 with 200 sperm and 40 barrels of oil. Sent home 218 barrels of sperm.

1863 *Falcon,* Brig, 159 tons, Capt. Holmes; owner or agent, John C. Osgood, to Atlantic, sailing August 31 and returning September 30, 1864, with 90 barrels of sperm, 20 barrels of whale-oil. Sent home 80 barrels of sperm.

1865 *Falcon,* Brig, 159 tons, Capt. Holmes; owner or agent, John C. Osgood, to Atlantic, sailing January 11 and returning October 5, 1866, with 127 barrels of sperm, 31 barrels of whale-oil. Sent home 288 barrels of sperm, 105 barrels of humpback oil and 470 pounds of whalebone.

Para, Schooner, 135 tons, Capt. Hussey; owner or agent, John C. Osgood, to Atlantic, sailing October 3 and returning October 20, 1867 with 140 barrels of whale-oil. Sent home 200 barrels of humpback oil.

1866 *Falcon,* Brig, 159 tons, Capt. Macy; owner or agent, John C. Osgood, to Atlantic, sailing November 26 and returning April 21, 1868 with 6 barrels of oil, apparently disposing of the cargo ahead of time.

William H. Shailer, Bark, 175 tons, Capt. Marshall; owner or agent, John C. Osgood to Atlantic, Oct. 17 to Oct. 13, 1867 with 100 sperm. This vessel was formerly in the African trade.

1867 *Para,* Brig, 135 tons, Capt. Worth; owner or agent, John C. Osgood, to Indian Ocean, sailing December 17 and returning May 16, 1871 with 760 barrels of sperm. Sent home 116 barrels of sperm. Altered from a schooner to a brig in 1867. Built at Wilmington, Delaware in 1861. Sold to Boston, 1871.

Said Bin Sultan, Bark, 235 tons, Capt. James W. Holmes; owner or agent, John C. Osgood, to Atlantic and Pacific Oceans; sailed June 13 and returned June 13, 1871 with 294 barrels of sperm and 149 barrels of whale-oil. Bought from Boston in 1867. Built at Newburyport in 1861. Sent home 410 barrels of sperm oil. Sold to Boston in 1872. This was Salem's last whaler.

William H. Shailer, Bark, 175 tons, Capt. Marshall; owner or agent, John C. Osgood, to Atlantic, sailing December 26 and sending home 243 barrels of sperm and 20 barrels of whale-oil. Condemned as unseaworthy and sold at Rio Janeiro November 6, 1869.

1868 *Falcon,* Brig, 159 tons, Capt. Richmond; owner or agent, John C. Osgood, to Atlantic, sailing on June 9 and returning on May 9, 1871 with 471 barrels of sperm and 4 barrels of whale-oil. Sent home 25 barrels of sperm. Sold to Boston 1871.

The late Colonel Lawrence Waters Jenkins of the Peabody Museum prepared a list of Salem whalers in 1908 which is pertinent to the record and sheds much light on Salem whaling:

	Rig	*Tons*	*Name*	*Voyages*	*Years*	*Notes*
1.	Brig		————*	1	1784	Captain Frost. 600 barrels of whale-oil taken to Boston in August.
2.	Ship		*Minerva*	1	1799-02	Sealing in the Antarctic.*
3.	Ship		*Rachel*	1	1803	Sealing in the Antarctic.*
4.	Brig		*Britannia*	1	1818	Sealing. Wrecked off Manchester.
5.	Ship		*General Knox*	1	1820	600 barrels of sea elephant oil.
6.	Brig		*Nancy*	1	1821	100 barrels of sea elephant oil.
			After revival of whaling industry, 1832.			
7.	Ship	304	*Bengal*	4	1832-40	
8.	Ship	...	*Catherine*	1	1832	Burnt at Hawaiian Islands, 1832.*
9.	Bark	295	*Cavalier*	1	1835-37	
10.	Bark	...	*Derby*	1	1837	Wrecked at Falkland Islands, 1838.
11.	Brig	...	*Eagle*	1	1837	
12.	Ship	262	*Eliza*	3	1833-41	Altered to a bark, 1838. Condemned 1843.
13.	Ship	397	*Elizabeth*	3	1836-44*	
14.	Brig	98	*Emeline*	2	1836-37	
15.	Ship	271	*Emerald*	5	1833-43	Altered to a bark, 1836. Lost at Madagascar, 1845.
16.	Brig	159	*Falcon*	5	1862-68	
17.	Sch.	89	*Franklin*	2	1836-37	
18.	Bark	262	*Henry*	2	1841-45	Lost in the South Pacific, 1847.
19.	Ship	275	*Izette*	6	1841-40*	
20.	Ship	395	*James Maury*	3	1833-41	
21.	Ship	280	*Lydia*	2	1835-37	
22.	Sch.	80	*Mac*	2	1836-37	
23.	Bark	268	*Malay*	2	1837-42	Lost in the Mozambique Channel in 1842.
24.	Bark	230	*Margaretta*	2	1851-52	
25.	Bark	216	*Messenger*	3	1854-60	One of the vessels of the "stone fleet" which sailed from New Bedford in 1861 and was sunk to block the harbors of Charleston & Savannah.
26.	Ship	325	*Mount Wollaston*	3	1836-40	
27.	Bark	249	*Palestine*	2	1835-39	
28.	Bark	...	*Pallas*	1	1832	Sailed for the Fiji Islands.
29.	Sch.	135	*Para*	4	1865-71	Altered to a brig.

	Rig	Tons	Name	Voyages	Years	Notes
30.	Bark	230	*Reaper*	4	1833-39	
31.	Bark	252	*Richard*	2	1835-37	Brig *Eagle* sailed as tender on second voyage. Lost at Montevideo in 1839.
32.	Bark	302	*Said Bin Sultan*	1	1867*	
33.	Ship	372	*Samuel Wright*	3	1833-39	Lost at Geography Bay, Australia, in 1840.
34.	Ship	365	*Sapphire*	3	1836-42	Foundered in the West Indies, 1843.
35.	Bark	258	*Statesman*	3	1836-42	Condemned in 1844.
36.	Bark	175	*William H. Shailer*	2	1866-67	

Available tonnage has been added to Mr. Jenkins' list.

1. *Captain Frost's brig has now been identified as the *General Knox.*
2. *Sealing was included in early whaling voyages.
8. *The destruction of *Catherine* is given by Starbuck as 1834.
13. **Elizabeth* made a third voyage from 1844-48.
19. **Izette*. Date is obviously a misprint. Should be 1831-1842.
32. **Said Bin Sultan's* voyage terminated 1871. Salem's last whaler.

Additions from Starbuck.

Brig *Polly* sailed in August of 1820.
Brig *Charles Doggett* made 1 voyage in 1833.
Ship *Clay* made 1 voyage in 1833.

CHAPTER

9

Merchant Service or Whaling Cruise?

The Skipper's on the quarter-deck
a-squinting at the sails,
When up aloft the lookout sights
a school of whales.
Now clear away the boats, my boys,
and after him we'll travel,
But if you get too near his fluke,
he'll kick you to the devil!!
J. C. Colcord's *Roll and Go.*

From the early years of its settlement in the seventeenth century, Salem was active in shipbuilding and sea-commerce, reaching its greatest maritime fame after the Revolution. During the Revolution Salem had fitted out 158 vessels as privateers. When peace came, these swift sailers were too competent for coastal trade and invited longer voyages, which became exploratory expeditions. The story is well known how the privateers of the Revolution, converted into merchantmen, helped to open up the markets of the world, bringing to Salem warehouses, cargoes of silk from India, tea from China, pepper from Sumatra, coffee from Arabia, spices from Batavia and hides from Africa.

At the time Salem's whaling efforts were at their peak, various other types of trade still flourished of interest to seamen and ship-owners. The South American hide trade was very lucrative as well as the new business of importing America's first gum overshoes from Para, Brazil. Trade of gum copal used in making varnish was active with Zanzibar and Madagascar off the coast of East Africa. There was the trade with the west coast of Africa in which Salem ships returned with ivory, palm oil, peanuts and gold dust. Salem also enjoyed a considerable trade to the Fiji Islands for bêche-dē-mer, the sea-cucumber delicacy esteemed in the Canton markets and exchanged there for tea, etc. After the peak of foreign commerce had been reached, alert merchants looked to replace it by instituting the whale fishery in the 1830's.

In view of the rivalry of marine pursuits within Salem itself, it is not unusual that only a comparatively few could be attracted to the difficult and, in most instances, unprofitable whaling voyages. Nevertheless the industry with its unusual aspects of challenge to masculine strength and skill had its devotees in Salem and the surrounding areas.

There was a pattern of life in Salem which had become a tradition. At fourteen a Salem boy sailed on his father's vessel. At twenty he had become a captain. At forty he retired to enjoy his wealth and leisure in an elegant home which reflected the success of his daring and achievements. Some hoped to make fortunes in whaling, or at least to advance to the rank of an officer as a step to success in the merchant service. In Salem, whaling was never more than an intriguing branch of the great shipping business. Whaling never secured the foothold which created whaling families such as existed in New Bedford, Nantucket and Fairhaven.

In these ports were circles of whaling people whose lives were completely wrapped around this industry, whose members comprised whaleship builders, officers, and those who did not go to sea but invested their money and became owners or stockholders in the enterprise. New Bedford became a network of marriages and inter-marriages in the whaling clans. The entire population of Nantucket once derived a livelihood from whaling. This type of thing existed in Salem in regard to the shipping industry. The ropewalks, ship carpenters, ship builders, suppliers of provisions, warehousing and even banking and insurance companies were interwoven with Salem's chief occupation.

For the requirements of a whaling cruise, the crew could be made up of men with little sea experience, or green hands, but the captain and officers were always experienced seamen and whalemen. If these were not available in Salem, they were secured from elsewhere as the log of the *Elizabeth* indicates on her voyage of 1836. The commander was Isaac G. Hedge of Barnstable, Cape Cod, and Henry G. Cotton the first officer came from Nantucket. Abner Lake of New Bedford was Second Officer and John Willson of Albany, New York, was Third Officer. In listing the crew for this voyage, the ages are given and they range from fifteen to thirty-six. While many of the men are from Salem, some are from other parts of New England, two youths from London, England.

Crew of the *Elizabeth*:

Gilman Lang, 33, Deerfield, Conn.
Abraham Anthony, 36, New Bedford, Mass.
Charles Nye, 26, Barnstable, Mass.
Joseph Manerell, 24, Middleton, Mass.
Jonathan Shirley, 23, Salem, Mass.
Hawens Hulling, 23, North Kingston, N.H.
James Kimball, 24, Salem
William Furness, 22, Boston
Moses Trask, 21, Salem
Israel M. Larrabee, 21, Durham, N.H.
Isaac Miles, 21, Limerick, Maine
William Holmes, 24, Boston
Robert Martin, 21, London, England
George Greaton, 19, London, England
Samuel T. Gremer, 19, Newfield, Me.
Benjamin Lund, 19, Merrimac, Mass.
Leander Doane, 19, Petersham, Mass.
William G. Lyman, 19, Columbia, N.H.
Royal E. Pulsifer, 18, Rockingham, N.H.

Edward Cloutman, 17, Salem, Mass.
George F. Gerrish, 17, Salem, Mass.
William G. Wright, 15, Salem, Mass.
Josiah Wyatt . . . "no proof." Not on board
J. M. Langley . . . "no proof"
Amboy Lippen, Cook, proofs wanting
Horace C. Gibson, steward, "proofs"

The log of the *Elizabeth* is particularly illuminating as even the duties of officers and crew are indicated. We find that the Starboard Quarter boat is within the scope of the Captain; Abraham Anthony is steerer; James Kimball is bow oarsman; Haw. Hulling is midship oarsman; William G. Wright is tub oarsman and Moses Trask or J. Langley are aft. Men are also named for the Larboard Quarter, Waist Boat, Besse boat and Moses Trask is listed as ship cooper, Haw. Hulling as carpenter.

This well-planned voyage of the *Elizabeth* lasted for three years, five months and 27 days. There were long periods of comparative rest on such a voyage which were punctuated by driving, disagreeable work. In general, duties on a whaler were not as exacting as those on a merchantman. In order to increase speed, a seaman had to climb the rigging of a merchant vessel a dozen times a day when the whim of the wind demanded a change in sail. A whaler never concerned itself with speed. A whale might be found as easily in one water as in another. A merchantman had to be kept spotless everywhere. Although a whaler's fo'c'sle might be foul, her deck was always kept immaculate until whales were caught, then the whale dominated the ship.

Having no particular destination toward which it must hurry, a whaler became the isolated home of its men, who watched and waited for the chance to earn something from the ocean itself. The whaler and its whalemen, declared Melville, were the true owners of the seas while other seamen enjoyed but a right of way through it. "Merchant ships are but extension bridges, armed ones are but floating forts; even pirates and privateers, though following the sea as highwaymen the road, they but plunder other ships, other fragments of the land like themselves, without seeking to draw their living from the bottomless deep itself." The whaleman alone ploughed the sea as his own special plantation. He knew the wet places of the world as a farmer knows his fields, and garnered the harvest of the whaling grounds.

Even those seamen who preferred the handsome merchantman to the square-sterned whalers, priding themselves on their seamanship and disdaining the duties of the whalemen, had to admit that the whaling service required a skill which was unique. A standard of *boatmanship* existed apart from regular seamanship which had no equal in marine pursuits. Whalemen were the most resourceful seamen this country ever produced. Perhaps no other maritime pursuit asked so much of a man, enduring unbelievable hardships, sailing for months on end without sight of land and gambling their lives in physical combat with the mightiest monsters since the age of the dinosaurs.

During the time the ship was on the whaling grounds, there were as many as five men on the look-out. Men were placed at each mast-head who were relieved every two hours; an officer was also on the fore-topgallant-yard so that there were four of the crew constantly on the watch from the most elevated parts of the ship. Generally there was a watch on deck and one below except when whales had already been taken, then all hands were employed in the trying-out.

Then at last came the cry from the mastheads of "Whale ahead. Thar she bloo-ww-ss!" At least two boats properly manned were lowered into the water within one minute flat of the order being given after sighting the whale. Only first class organization and discipline could achieve this feat. The conflict of a few men in a boat dancing about on top of the waves next to a whale three times its own length and as high as a wall gave the work a pace absent in most occupations. Alert thinking, dexterous management of the harpoon line, the boat, and above all the daring in placing the boat alongside of the whale for the high moment of the kill, when the whale is conscious of its jeopardy and plans his counter-attack, put whaling in a class by itself.

Few sports or occupations connected with the sea could afford more drama and thrills than was present in the whaleboats. It was a real man's life for those who were happy roving and roaming all over the oceans and could bear the physical limitations of the ship for endless months, the challenge of the whale hunt and the cutting-in and trying-out marathon. No one but an exceptional sailor could have stood it or survived the exposure to scurvy, famine and terrible danger.

A green hand could ship on a whaler and by trial and tribulation learn something about whaling. A whaling crew, however, included many skilled men such as riggers, coopers and mechanics. Boys who longed for a life on the sea felt they could get a start in whaling and advance to the merchant service. Between the years 1815 and 1840, the standards of American seamanship were high. "No man" says Professor Samuel Eliot Morison in his *Maritime History of Massachusetts* "could be rated an able seaman until he became an expert in the beautiful splicing, seizing, parceling, gaffing, pointing and worming and serving which was included in the old-time art of rigging. An ordinary seaman was expected to hand, reef and steer . . . to be able to reeve all the studdingsail gear and to set a topgallant or royal studdingsail out of the top; to loose and furl a royal, and a small topgallant-sail or flying jib, and perhaps also to send down or cross a royal yard."

These were the triumphant years of Yankee seamen with a large number of Englishmen and Scandinavians serving in Massachusetts vessels. In the merchant service, seamen wore attractive uniforms—shiny black tarpaulin hat, red-checked shirt, blue bellmouthed dungaree trousers, navy blue pea-jacket or watch-coat when he was on board ship. For shore leave there was a black-ribboned hat, black silk handkerchief in a neat sailor's knot around the neck, white duck trousers and black pumps. Little wonder that such fine-feathered birds of the sea looked down upon the drab whalemen sailing on their slow and cumbersome ship.

It was an era when a boy had the chance to rise on a small vessel to the rank of an officer. Wages from 1815 to 1840 were $8 a month for boys, $10 for ordinary seamen and $12 a month for able seamen. This scale of pay applied to service in the merchant marine. Whalemen received no wages. Straight wages would not have proved workable. Fishermen from ancient times to the present have shared the work and the profits of the catch. Thus the *lay* or share system was the only workable scheme for remuneration on a whaling voyage. A chance to share in the spoils furnished the maximum of incentive. Sometimes the division was not fair, but a *full* ship—one loaded with oil on return—generally saw all hands well paid. As bad luck voyages often occurred, the whaleman received little return for his time and work. The point is that all hands had to share in the gamble of a good or poor voyage.

Yet few branches of human endeavor at sea were more dramatic or provided more poignant instances of human laughter and human tears than whaling. The very nature of the whale, its awesome bulk, its power, and even its seemingly intelligent malignancy when cornered, made whaling men constantly aware of the elemental forces they opposed. No merchantman knew the excitement and thrill of the chase, the nerve-tingling shock of the 'Nantucket sleigh-ride,' the satisfaction of the capture, and, in fine, the all-out hazards of that calling which was both sport and livelihood to those who were part of it.

CHAPTER

10

George Dodge of Salem Ships on a Whaler

"The Mariner that on smooth waves doth glide
Sings merrily and steers his Barque with ease,
As if he had command of wind and tide,
And now becomes great Master of the seas;
But suddenly a storm spoils all the sport,
And makes him long for a more quiet port,
Which 'gainst all adverse winds may serve for fort."

From *Contemplations*
Anne Dudley Bradstreet (1612-1672)
of Salem, America's Earliest Poet

George Dodge was twenty-three years old. All of his life he was part of the sea, smelling its salty spray, wandering about the wharves of Salem, watching the frigates, merchantmen and whalers recruiting for long voyages to distant lands, watching the sailor return, tanned, bearded, worldly, sometimes in strange clothing and often as not with a parrot on his shoulder. As a little boy he listened enthralled to the tales of adventure. And he knew that one day he, too, would be a mariner on the breast of the ocean. His choice would be the most dangerous of callings, a whaling cruise.

"A whaler's life" wrote Chatterton in his *Whalers and Whaling* "was likely to appeal to three classes of men: those who had been compelled to leave the land to avoid jail or starvation, those who thought they were going to see the world and gain adventures, and those who were determined to work their way up till they owned a whaling ship of their own. It was only the latter who ever stuck to the occupation for voyage after voyage."

Shipping agents engaged crews by promising large profits, adventure, and a good sea life. In *A History of the Whale Fishery,* the author James T. Jenkins includes a typical advertisement broadcast in the eastern seaboard states.

WANTED LANDSMEN: 1,000 stout young men, Americans, wanted for the fleet of whaleships now fitting out for the North and South Pacific Fisheries. Extra chances go to Coopers, Carpenters and blacksmiths. None but industrious young men with good recommendations taken. Such will have superior chances for advancement. Outfits to the amount of $75.00 furnished to each individual before proceeding to sea.

Persons desirous to avail themselves of the present splendid opportunity to see the world and at the same time acquire a profitable business will do well to make early application to the undersigned.

Gullible New England and New York boys who craved adventure or longed to know something of the wide world, read the advertisements and circulars—and signed up. "It was the custom" recalled Dodge many years after his own experience "to have an agent in Boston to ship green hands from the country. They told fine stories about whaling profits and what a pleasant voyage it was and held out every inducement." Many are the stories told *about* these whaling agents. Professor Morison includes a classic in his *Maritime History of Massachusetts* which bears repeating. "A well-known Boston agent, after describing to a Maine ploughboy the imaginary joys of this glorious profession, concluded confidentially: 'Now, Hiram, I'll be *honest* with yer. When yer out in the boats chasin' whales, yer git yer mince-pie *cold!*'" George Dodge of Salem was a New

England boy who could not resist the alluring invitation of far-away places and life at sea. In 1831 he signed up and shipped aboard a whaler.

George Dodge left a good account of his whaling voyage faithfully committed to a notebook. A half century later, fearing it would be lost to posterity, he took it to the *Salem Gazette,* in 1882. The presses printed his story in a tiny thirty-page pamphlet 4″ x 7″. Only one copy of it was ever found and it floated about among the flotsam and jetsam of literature in the old archives of the Essex Institute. One day I received it in a letter from Mrs. MacSwiggan who was associated with the Institute.

I felt indebted to her eagle eye for spotting and rescuing it for inclusion in this Salem story. Aside from the whaling logs, here was the only authentic whaling account in print by a Salem man we ever were able to locate in the four years of research devoted to this yarn. George Dodge would be immensely edified if he only knew to what good purpose his notes have been put. Curious about him, we looked up the old *Salem Directories.* These showed that George A. Dodge, mariner, lived at a house at 8 Conant Street in 1859, a Mr. Dodge boarding at 29 Union Street, next door to Hawthorne's birthplace, in 1861; listed as a clothing store owner and grocery also at Derby Street in 1879, and living at a house at Washington Square in 1882, the date of the pamphlet. In the absence of other information, it is felt that this person is the George Dodge of the whaling yarn.

Leaving Salem in 1831, Dodge went to Boston and from there to Nantucket, where he became one of a crew of thirty men on board the *Baltic,* commanded by Captain Chadwick. The *Baltic* shipped four mates and four harpooners. It was customary for the captain to make a speech to the men after the ship was under way. He laid down the rules of the voyage and told the crew they came to work hard, catch whales and get home in good time. The language the whaling skipper used was different from that employed in the counting rooms of the owners. With many a round oath, he insisted that under no circumstances would he allow swearing on board his ship.

On the first night out, after the excitement of departure had subsided, the greenhorns were most likely prostrate with seasickness, wondering disconsolately why or how they had been such fools as to come aboard for an adventure of this kind. Most of them anticipated enjoying financial security for a number of years by shipping on a whaler. It would be such a gay life with no worries about paying board or bills. With the dawn came also the realization that this whaler was no second home at all!

On the first day out, it was customary to divide the men into watches and boat-crews and some logkeepers made a complete record of this in the ship's journal. The officers tried to get the ship into full working order and in their efforts at organization often bullied the men. Dodge mentions the fact that the "mate had a cat-o-nine tails and flogged the men with it." There was the eternal practicing of lowering the boats. Some of the boys had to learn how to row, to handle the long oars, to sail and to use paddles. In the pursuit of a whale both rowing and sailing the whaleboat increased speed and one must be handy with the paddles in stealthily nearing the monster so as not to frighten him. His eyesight may be poor, the officers warned, but his hearing is very keen. After a few days of disillusion, the crew—from many places and of different nationalities—became reconciled to ship food and to each other and settled down to what a trained seaman called a "sailor's horror"—a cruise on a whaler.

Many a disappointed would-be whaleman deserted a "floating hell" in a distant port and became a beachcomber on some balmy island in the Pacific, or, as whalemen called it—a "lone man." Sometimes the geography was not so pleasant. An old ballad with salty humor tells of a whaleman's plight in the frozen northlands:

"He shipped me on board of a whaler
Bound to the Arctic seas
Where the cold wind blows
And it frosts and snows
And Jamaicy rum would freeze
And worst of all
I had no clothes
I spent all of my money on shore
Oh, it is then I wish
That I was dead
Or back with the girls once more."

The *Baltic* under Captain Chadwick, who was a real old salt, was out a month before the first whale was seen. All was noise and confusion after the glad cry from the masthead of "There she blows! blooooows! There she *breaches!*" In the high spirits of his play, a whale *breaches* or leaps far into the air, landing on his back with a blow which resounds for miles around. "The boats were made ready to lower," recalled Dodge, "and everything prepared for the chase. The whale knows when he has been spotted. He sinks and stays under water one hour by the watch."

How the whale submerges or "sounds" remains his own secret. Dodge's first whale sank and stayed under water an hour, then rose to the surface and spouted once a minute for sixty minutes when he sounded again. Some observers have written on "dancing whales" and "diving whales." The Humpback Whale dives deeply then shoots for the surface at a high speed, hurtling his huge body into the air like a rocket. The Sperm is a deep diver when he sounds and heaven help a vessel in his way when he re-appears!

All hands were in high glee over the thoughts of beginning the pursuit for which they had shipped from port almost a month ago.

"Keep the ship close to the wind," said the captain "and get the boats in readiness to lower as soon as he rises, for he is a large whale and by his appearance looks as though he might be a hard customer to handle. Let every man do his duty and capture the monster . . . There she blows! close to the ship. Lower away your boats, boys!"

"When he came to the surface, the excitement was beyond description among the green hands who had never seen a whale. The whale being near enough, the Captain gave orders to take in the light sails, haul up the courses, haul aback the main yard and lower down the boats—all under one breath. The scene lowering the boats beggared description. It was no easy matter for a greenhorn to clamber down the sides of a ship in a heavy sea." We might add here that an equally athletic feat was to return to the whaler's deck by clambering up her side from a boat on the open sea, capricious as old Neptune is with great swells which now lift the boat high up towards the bulwarks, and in a second drop it half way down to the keel or bottom.

"When all the boats were down, the race began with each officer being eager to get alongside the whale first. Every man was stretching every nerve to capture the prize. Every time the whaleboats drew near, the whale went down and came up in another place. In this way four or five hours were consumed in rowing."

"Pull away, boys!" says the mate. "We will have her the next time she spouts. She is an ugly customer, but we must take her at all hazards."

The backs of the rowers were to the whale and the men all faced aft of the boat. The confidence of the crew in the judgment of the boatheader was important and they had to have discipline. The boatheader steered by using a tiller, or a 28-foot oar, while the harpooner pulled the forward oar which was 14 feet long until they were right next to the whale. During the rowing, the boatheader was the only one who could see where they were going and how near they were to the whale. He gave his orders accordingly. It was bad whaling form to glance around and the crew first knew of their nearness to the prize when the mate shouted to the bow-oar, or harpooner, to "Stand up and let him have it."

It is only *after* the whale has been fastened by a toss of the harpoon, that the harpooner takes up his role as boatsteerer. Up to the time of harpooning, he is at the oars. When he gets the command to dart the harpoon, he ships his oar, gets up his iron, and tosses it. A second harpoon is handy and he hurls it if there's time or necessity. After the whale is fastened, he moves aft to take over the steering oar and to tend to the towline which is generally 100 fathoms long. The line is fixed to the harpoon shank and threaded through a chock in the bow. It is the duty of the boatheader to look after the line until he is relieved by the boatsteerer. The boatheader then moves forward to kill the whale. Two lances are there and he throws both. Seldom if ever, was a whale dispatched by the toss of the harpoon. Sometimes a boatheader darted his lance fifty or sixty feet, and it would sink five feet of iron into the whale.

The vision of a muscular young whaleman poised gracefully in the whaleboat's bow as he tosses his slender weapon into the prize is pleasing but not true. The hand harpoon was no light Zulu spear. Actually, it was a clumsy, heavy, unwieldy instrument and it required great skill to balance and much strength to hurl it. The harpoon was usually about eleven feet long and as thick as a man's wrist. Its hickory shaft was eight feet and its iron shank was three feet. The shank made from a young sapling could bend double without breaking. Each harpoon had the name of the ship and its master stamped on it. Now, the lance, or spear, with which the whale was killed, had a ten-foot shaft, a five-foot shank and a razor-edged head to inflict the death wound. Sometimes the lance is referred to as an "iron." Dodge notes that the mate ordered the harpooner to stand up and throw his harpoon as quickly as they got within darting distance. When the whale is first harpooned he takes violent plunges in re-acting to the painful shock of attack. He throws his dangerous tail or flukes high in the air and lashes the sea into foam which obscures the men and threatens to overwhelm them. Then begins that tearing along the surface of the water, towing the whaleboat after him, which is known as the "Nantucket sleigh-ride." The men with their oars *peaked,* ready for use, resting at an acute angle with the gunwale have all they can do to stay in the boat as it goes thudding over the waves, throwing spray to port and starboard.

"After a whale is speared" wrote Dodge "he races through the water with the whaleboat in tow. A sleigh-ride is no comparison. A fast whale hits about twenty-five miles an hour." Dodge's estimate is a little high. The speed is somewhat nearer fifteen miles an hour. Still there was nothing comparable in the world of sports until the coming of the fast motor-boat days. "Sometimes" Dodge's narrative continues "a whale rounds out two or three hundred fathoms of line. When the whale has been exhausted by his run, the pursuers close in for the kill. When the harpoon fastens to a whale," warns Dodge, "you have to watch that he doesn't come up under the boat." When a whale achieves this counter-attack, the whaleboat is knocked twenty feet into the air. A flip of the colossal tail has smashed whaleboats to matchsticks. When he shakes this tremendous weapon, it cracks like a magnified whip, and resounds to the distance of three or four miles.

Dodge relates a near tragedy. The mate told the harpoonman to stand up "And when I tell you to dart, give it to her solid!" He obeyed the commands of the officer and sent the weapon chock to the hilt. This so enraged the victim that it struck a beeline for the boat. The officer's next command was to his men. "Stern all, boys, for dear life, or he will knock us all into the briny deep to make food for the sharks!" "Stern all" meant reversing the oars which backs the boat away from the startled beast. The officer's next cry struck terror into every boy's heart. Backing up with all their strength, through the roar of the great splashing they heard the words: "He is right on to us, boys, with his huge mouth open, ready to swallow us up, or send us flying into the air!" By some miracle, the cavernous mouth was avoided. The crisis passed. The whale began his run. The men overtook him. The lanceman hurled his weapon and the prize was won. The cessation of danger, the great relief which bathed every man in the whaleboats, and the ultimate conquest returned the men to the mother ship in jubilant spirits.

"The fourth mate" Dodge remembered "was the lucky man, getting the first iron into the whale, much to the mortification of the chief mate who was very anxious to get the first whale. The number of whales each officer gets is recorded in the ship's book and the one getting the highest number is made a hero of when the ship gets home and considered a greater man than the President of the United States. To get a wife in Nantucket you had to double Cape Horn three times, fasten to a whale, and do everything else appertaining to a whaling voyage."

"Everything else" included the frightful experience of capturing and killing the whale. Dodge described his first whale as looking upon approach "like a large ocean steamer paddling along with his fins and flukes." The creature upon being attacked changes his role from the hunted to the attacker, using every defense at his command. The flukes of the Sperm Whale is his deadliest weapon. It was by no means unusual or an isolated incident, Mr. E. K. Chatterton tells us, in *Whalers & Whaling* for the harpooner as he stood there in the bow of the boat to be killed by the Cachalot's tail sweeping through the air and cutting the boat off clean to the water's edge. It was characteristic of some Sperm Whales that they preferred to attack with their jaws rather than with their tail and would rush open-mouthed against the boat in an effort to crush it with their teeth, for this purpose turning and swimming on the back, a horrifying sight.

Also to be taken into consideration is the whale's family and friends. A Sperm seldom travels alone unless he is an aged bull.

A group of twenty-five to fifty Sperm Whales usually comprises cows, calves, and at least one very large fierce bull who acts as escort or leader of the Sperm pack, school or shoal. A violent battle takes place when a young bachelor bull decides to acquire a family. He challenges the family head or schoolmaster bull. The victor of the show of strength takes the family while the defeated mammal moves on in search of another group. Young, unattached males in a group are called a "pod." They are as frisky and boisterous as college boys on a spree. A "body" of whales is formed when two or more schools cruise in company. They are frequently seen in awe-inspiring numbers when migrating or moving to a definite locality. The males are more desirable as catches for they are the greatest oil producers.

An angry, wounded Sperm Whale becomes a dangerous enemy. He has been known to bite a whaleboat in half with his great jaws and to sink ships. Probably some vessels which have disappeared at sea were wrecked by this submarine contrived by Neptune. There was the case of the American whaler *Essex* of Nantucket in 1820. Rammed twice in a counterattack by an avenging Sperm Whale, the *Essex* sank in ten minutes. Her crew drifted in the open whaleboats for two days until those who had survived the ordeal were at last picked up by passing ships. Another whale is known to have freed a mate by biting off the harpoon line. The instinct act of heroism cost the whale her life. She became entangled in the line and met death by the whaleman's lance.

The Leviathan has a particular vulnerable spot, like Achilles' heel. It is just behind either pectoral fin. He is also rendered fatally vulnerable by the structure of his circulatory system. When a land animal is wounded, the flow of blood is shut off by valves. One of the peculiarities of the whale is the non-valvular structure of his blood vessels. When he is pierced by so small a point as harpoon or lance, a deadly drain commences upon his whole arterial system which leads eventually to exhaustion as more wounds are inflicted and energy is destroyed in his fight

Whaling off-shore on Long Island, New York. Drawn by W. P. Bodfish. Author's collection.

"The Capture" From the Allan Forbes Collection. Lithograph of Sperm Whaling after drawings by A. Van Best and R. S. Gifford, corrected by Benjamin Russell. Endicott & Co., New York, 1862.

"In a Flurry" From the Allan Forbes Collection. Lithograph by N. Currier, New York, 1852.

This Charter party of Affreightment made & fully
Concluded upon the Seventeenth day of February Anno Domini
1743/4 Between Israel Nichols Mariner Thomas Andrews
Yeoman & Elisha Levitt Blacksmith all of Hingham in the
County of Suffolk & province of the Massachusetts Bay in New
England & owners of the Good Sloop called the Betty & Ruth of
the Burthen of fifty Tons or thereabouts now lying in the Har-
bour of Boston whereof Israel Nichols was lately master on the
one part and Timothy & Ebenezer Prout both of Boston in
the County of Suffolk & province afores.d Merchants of the
other part Wittnesseth That the said Israel Nichols & Comp.a
afores.d for the Consideration herein after mention'd hath Granted
& lett to freight unto the said Tim.o & Eben.r Prout afores.d &
they have accordingly hired of the said Israel Nichols & Comp.a
afores.d the Hull of the said Sloop Betty & Ruth with all her
tackell apparell furniture Boat, Stores, & appurtenances &
Whatsoever unto her Belonging for a Voyage with her to be
made from Boston aforesaid on Whaling On the Banks
to ye Southward as Customary for to Catch Whales ——
for the space of Three or four Months More or less & from thence
to Boston again & the said Israel Nichols & Comp.a Doth Co-
venant & agree to & with the said Tim.o & Eben.r Prouts That
the s.d Sloop shall be well & sufficiently furnished with all things
Necessary for the sea during the Voyage afores.d at the propper
Cost & Charge of the aboves.d Israel Nichols & Comp.a Excepting
Ballasting Victualling Maning & all Port Charges ——
& the said Timo. & Eben.r Prout do by these presents Covenant
& agree to & with the s.d Israel Nichols, Tho.s Andrews Elisha
Levitt to pay or Cause to be paid unto the s.d Israel Nichols and
Comp.a In good Bills of Creditt of New England the Sum of

Old whaling agreement date 1743/4, Boston, Massachusetts. Collection of H. Harrison Huster, N.Y. Member of Nautical Research Guild.

Vessel rigs. Drawn by Charles G. Davis, naval architect and famous ship modeler of t
Frigate Constitution, *the* Sea Witch, *the Whaler* Sunbeam *and others.*

Ship Emerald *of Salem. Water color by M. Macpherson of original by G. M. White.*

hip Bengal *of Salem in company with the* Foxhound *of England, Pacific Ocean haling grounds.*

A whaling station on the California coast by Frenzeny. From Harper's Weekly (Journal of Civilization), New York, June 23, 1877.

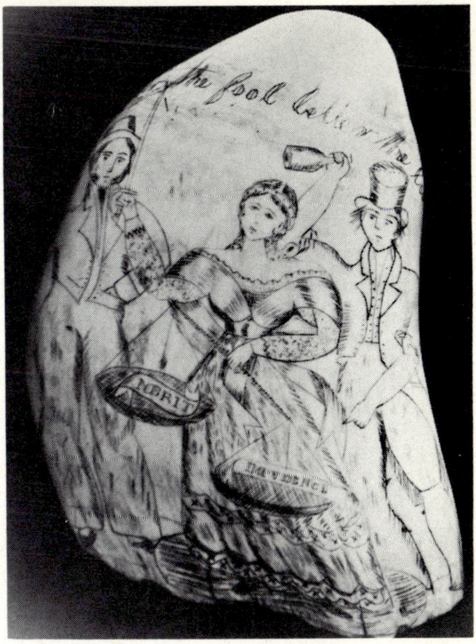

Scrimshaw on whale's tooth, front and reverse. Homeward Bound.

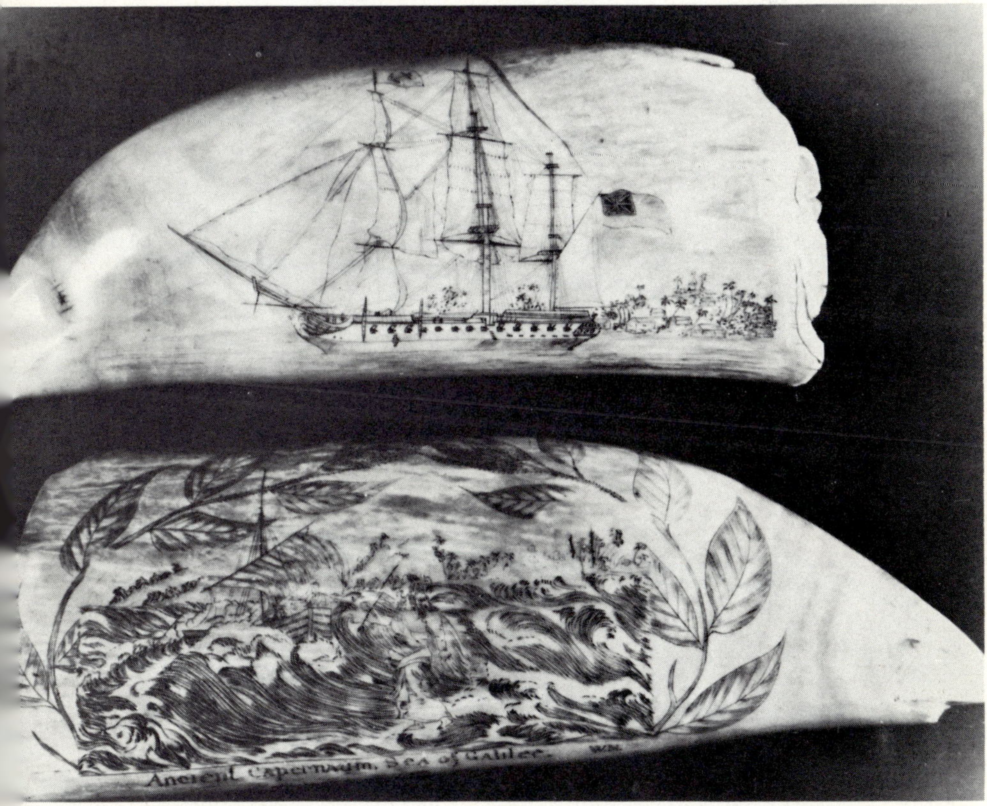

Scrimshaw of whaler flying the British Jack in ancient Capernum, Sea of Galilee. H. Harrison Huster Collection.

Charles W. Morgan, *last American whaleship. Marine Museum, Mystic, Connecticut. Photograph by Herbert H. Corey.*

Model of Whaling Brig Viola. *H. Harrison Huster Collection.*

South Sea whalers boiling blubber by O. W. Brierly. Illustrated London News, July 1, 1876. J. Geenaway.

EASTERN
EUROPE
AFRICA
ARABIA
BARBARY
SAHARA or GREAT DESERT
UPPER GUINEA
CONGO
LOWER GUINEA
UNKNOWN PARTS
ABYSSINIA
MADAGASCAR
ATLANTIC OCEAN
SOUTH ATLANTIC OCEAN
SOUTH
ANTAR
West from Greenwich
East from Greenwich
CAPE of GOOD HOPE
S^t. Helena
Tombuctoo
Mecca
Socotora

The Eastern Hemisphere as depicted in world map dated 1840. Published by Henry Teesdale & Co. London, England.

THE SAILING FROM NEW BEDFORD, NOV. 15, 1861, OF FIFTEEN OLD WHALERS LOADED WITH STONES, TO BE SUNK IN THE CHANNELS OF THE HARBORS OF CHARLESTON AND SAVANNAH TO PREVENT BLOCKADE RUNNING.

From a lithograph by L. Prang & Co., 1862, in the Allan Forbes Collection.

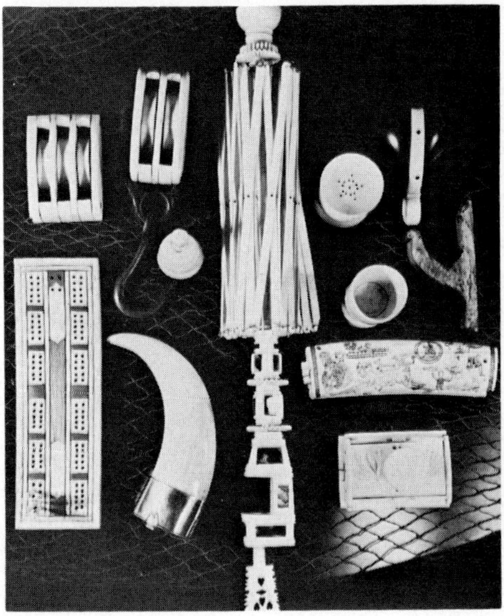

Clifford W. Ashley (The Yankee Whaler) *mentions some 60 different items of Scrimshaw in four categories: ornamental, utilitarian, personal trinkets and play.*

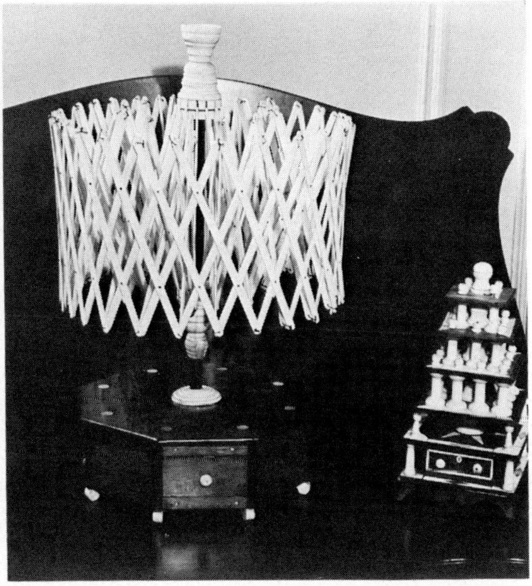

Whalebone swift, used in winding yarn for knitting and spinning, required a high degree of mechanical craftsmanship. H. Harrison Huster Scrimshaw Collection.

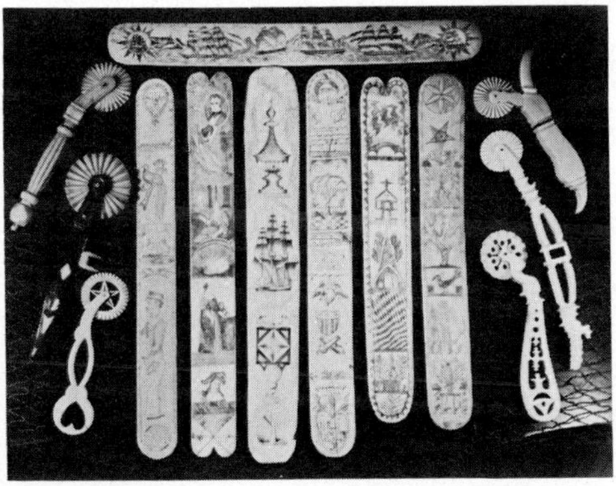

Whalebone busks, frontal center stay of a woman's corset. Jagging or crimping wheels for marking the edges of pies or tarts.

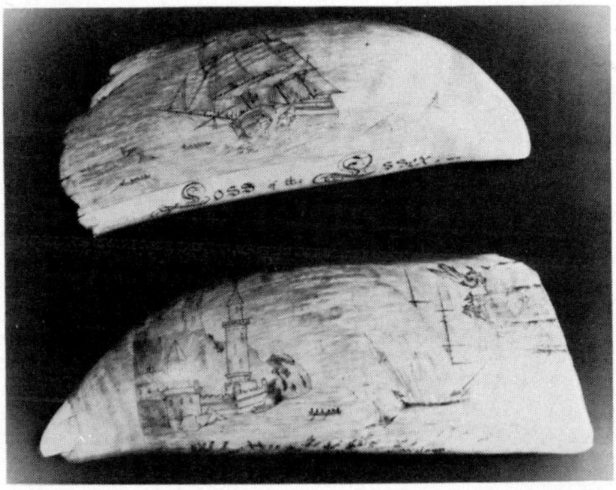

"My God! Mr. Chace, what is the matter?" said Captain Pollard. I answered: "We have been stove by a whale!" Narrative of the Shipwreck of the Whale Ship Essex of Nantucket *by First Mate, Owen Chace, New York, 1821. A Sperm Whale sank the* Essex *in ten minutes. Twenty survivors drifted in three rowboats for three days before their rescue. H. Harrison Huster Scrimshaw Collection.*

Model of a Fiji Island temple. Kindness of the Peabody Museum.

Cannibal bowl from the Solomon Islands. Kindness of the Peabody Museum.

Moorea, Tahiti—Peak over Cook's Bay.

Tahitian Vahine doing her daily laundry.

Couple dancing in costumes of the Island of Bora Bora, Tahiti. Kindness of Air France: Tahitian Holiday.

Rajendra Dutte, a merchant of Calcutta. Life-size portrait figure. In Peabody Museum, Salem, Massachusetts.

Life-size figure of Canton merchant. Head and hands were carved by Salem's Samuel McIntire. In Peabody Museum of Salem.

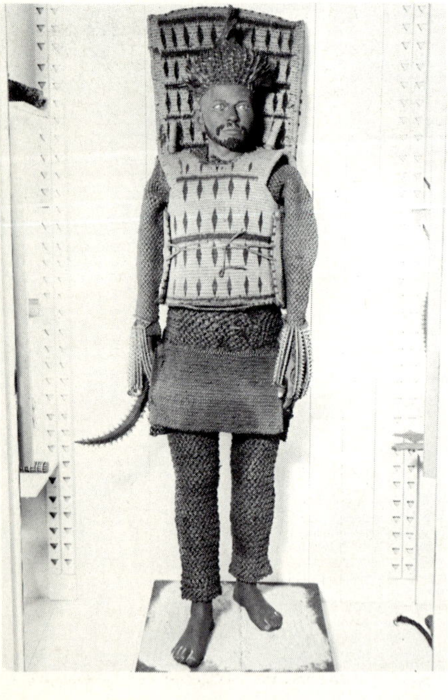

Gilbert Island warrior in coconut fibre armor with rare gauntlets set with shark's teeth. Life-size figure in Peabody Museum of Salem.

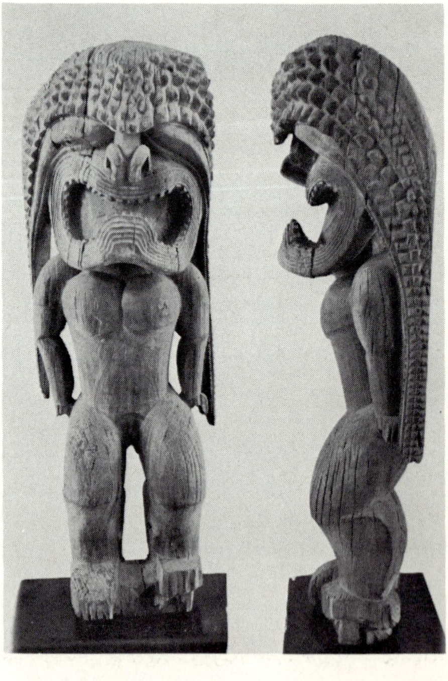

Hawaiian idol, six feet tall, one of many found by white men in a sacred island enclosure. Only three are in existence: British Museum, London; Bishop Museum, Honolulu; Peabody Museum, Salem.

for life. When he sinks to the protecting depths of the ocean, his danger is heightened by the unusual pressure of the water while "his life may be said to pour from him in incessant streams." Yet, says Melville, so vast is the quantity of blood in him, and so distant and numerous its interior fountains, that he will keep thus bleeding and bleeding for a considerable period.

Each dart of the lance opens new faucets, and the sea surrounding the beleaguered monster crimsons as the red tides pour from all sides like brooks down a hill, changing the white brine into a troubled whirlpool. But not until the "life" of the whale is struck, the spot behind its pectoral fins, is the victory over him complete. Then occurs that unspeakable climax of the hunt called the whale's "flurry." In desperate agitation, he wallows in the seething red pond and wraps himself completely in the boiling spray he has created until his own vastness is lost to the view of the fascinated men in the whaleboat. Swimming furiously in rapidly narrowing circles, the flurry abates and once more he rolls into view while the breath comes hard and sharp. At last, with excitement at fever pitch, the whale spouts scarlet. Gushing in a terrifying fountain, the clotted red gore signalizes that the big heart has burst.

"After making the whale spout blood" Dodge explains "you keep at a distance and let him die which takes about an hour. They live until all the blood is spouted up. When they are about to give up the ghost, they take a circle around and heading toward the sun—they expire. The ship runs up and takes the whale alongside." Then he adds for the sake of accuracy "We got three whales today." The Sperm and Right Whales are the only ones which float when dead. Air has to be pumped into the others to keep them afloat.

Sometimes a whale managed to escape capture, trailing as many as seven harpoon lines after him and sometimes it was the whale who won in the throw of the dice by Life and Death. Dodge relates the failure of a harpoonman at the high moment of the ordered thrust. "Owing to some mismanagement the harpoonman missed the whale. The mate talked to him very severely and hard words passed between them. Finally, the harpoonman told the mate to put him alongside of the whale again, and he would strike her, if she knocked him into hell the next minute. They pulled to the whale again, going to the windward when they should have gone to the leeward so as to have been off easy. The harpoonman struck the whale, the whale stove the boat and swept them all into the deep blue sea . . . The harpoonman was lost."

For the loss of a whale, it was customary to blame the harpoonman for his lack of skill or courage. There was for instance the time when two whalers were enjoying a get-together or *Gam.* Whales were raised in the midst of the fun, the boats were lowered and the crews of the two whalers contended for the prize. Just as the harpooner of Dodge's ship, the *Baltic,* was about to thrust his harpoon, the harpooner from the other whaler darted his harpoon over the bow of the boat—and secured the whale. Great was the chagrin of the losing team, as they turned about and pulled back to the *Baltic* without the whale. The prize lost represented 100 barrels of oil worth $3,000. Dodge complains with some bitterness that they lost the whale to the other crew "because of a chicken-hearted harpoonman. Many were the curses on the poor fellow's head for missing such a good chance for there was no excuse. The whale lay like a log on the water and there was but a little sea at the time."

Perhaps nine times out of ten there *was* excuse for a harpooning failure. That the lost art of hand harpooning a whale was a marvel of unequal distribution of responsibility was rarely suspected or admitted by the devotees of the big game hunt of the seas. A word or two in behalf of the old-time harpooner. The harpooner, as we have seen, pulled the foremost or harpooneer-oar, and by the usage of the whale fishery was expected to set an example for the crew of superhuman activity by pulling on the oars with all of his strength.

As if this were not enough of an exertion for one soon supposed to perform a feat of delicate skill and daring, when he reached the whale, the harpooner was also expected by whaling tradition to shout to the top of his lungs all sorts of exhortations to the crew as they rowed. This was a whaling jargon, sometimes obscene, of urging, cajoling, bullying the men to speed up the race toward the whale before he dived to the depths to re-appear a half hour or so later somewhere else. "Roar and pull, my thunderbolts!" cried the little "King-Post" of *Moby Dick*. "Beach me, beach me on their black backs, boys, only do that for me, and I'll sign over to you my Martha's Vineyard plantation, boys, including wife and children, boys. Lay me on—lay me on! O Lord, Lord!" etc.

Straining and yelling and nigh unto exhaustion, with his back to the whale, the harpooner suddenly hears the exciting command: "Stand up and give it to him solid!" He drops and secures his oar, seizes his harpoon from the crotch, and with what strength he can muster pitches it into the whale. His chances of success are about five out of fifty. The whaler cruises on without the whale and the back of the log book lists the numbers of whales lost. "No wonder" lamented Melville "that to many shipowners, whaling is but a losing concern; for it is the harpooneer that makes the voyage, and if you take the breath out of his body, how can you expect to find it there when most wanted!"

If the harpooner be successful, then arises another critical moment before the headsman or temporary boatsteerer can dispatch the whale. Just as the wounded animal begins to lurch forward to begin his wild, desperate run, the boatheader, or lanceman, (who is the chief officer of the little pursuit craft), and the harpooner stand up at the same moment in the ends of the 28-foot whaleboat and begin to run fore and aft at the risk of falling overboard and upsetting the boat. This exhibition of agility, in strict keeping with the rules of the whaling game, allows the headsman, who will throw the lance to kill the whale, to take his proper place in the bow of the boat. The harpooner's race to the stern is to take over the abandoned steering oar. The chances of the whale to make a clean get-away are very good.

George Dodge served on the *Baltic,* Captain Chadwick, out of Nantucket on a cruise which lasted 44 months. His share of the profits was $125.00, according to the careful calculations of the purser. Returning to his home in Salem and reflecting upon it, Dodge felt that he had made a good voyage and had learned much. He possessed in lieu of a full purse a personal knowledge of far away places, of the ways of the whale and its pursuit, all there was to learn about life on board a greasy whaler, about officers and crew, about human nature, of elation and despair, of a code of behaviour, of ethics, of brutality. Indeed it constituted a wealth of knowledge.

By historical coincidence, the year Dodge's sea journal was published, was the year which marked the death of Henry Wadsworth Longfellow who had an affinity for the sea. The lines of the white-maned poet of Cambridge, Massachusetts, which he called *My Lost Youth,* captured the lure of the sea, the longing of youth for adventure, and the answer of the sea in good measure.

My Lost Youth

"Often I think of the beautiful town
That is seated by the sea,
Often in thought go up and down
The pleasant streets of that dear old town,
And my youth comes back to me.
And a verse of a Lapland song
Is haunting my memory still:
'A boy's will is the wind's will,
And the thoughts of youth are long, long thoughts.'

I remember the black wharves and the slips,
And the sea-tides tossing free,
And the beauty and mystery of the ships,
And Spanish sailors with bearded lips,
And the magic of the sea.
And the voice of that wayward song
Is singing and saying still:
'A boy's will is the wind's will,
And the thoughts of youth are long, long thoughts.'

Strange to me now are the forms I meet
When I visit the dear old town;
But the native air is pure and sweet,
And the trees that o'ershadow each well-known street,
As they balance up and down,
Are singing the beautiful song,
Are sighing and whispering still:
'A boy's will is the wind's will,
And the thoughts of youth are long, long thoughts.' "

Henry Wadsworth Longfellow
(1807-1882)

CHAPTER

11

The Try-Works: From Blubber to Oil

Whales in the sea
God's voice obey.
—New England Primer

A large whale taken alongside the ship usually reaches from the ship's fore-chains to her mizzen—about 75 feet. Preparations get under way to cut-in the blubber by large tackle and fall made fast around the head of the mast. Staging is built over the side of the vessel for the officers to stand on while cutting-in. The harpooner's duty is to go on top of the slippery whale to hook on to the first piece. It is difficult and dangerous when there is a heavy sea and naturally falls to the lot of the harpooner from whom so many heroic things are expected. The sharks, who follow the whale blood, now anxious to get at the carcass heighten the harpooner's peril. When the blubber is completely removed from the body of the whale, the head is separated and hoisted up at the side of the ship. Some writers say the head is first severed and hoisted on board. At any rate, a bucket is set into the head to get the oil and it is hoisted out with a tackle. In this way some 15 to 20 barrels of oil are taken out of a Sperm Whale's head. This head matter needs no trying out except heating to keep it from turning rancid. But let us consider the try-works by which the whale is reduced from monarch of the seas to oil for the lamps of gay bridals and for the scholar poring over his books at midnight.

No feature was more characteristic of the American whaler than the great ovens or try-works in which the blubber was converted into oil. The try-works were of the most solid masonry imaginable which joined with oak and hemp in constituting the completed whaleship. These works were situated between the foremast and mainmast which was the most roomy part of the deck. The mass of brick and mortar occupied usually some ten feet by eight feet square and five feet in height. The timbers beneath this great mass were of unusual strength which fitted them for the responsibility of bearing up the most important equipment of the whaler. The foundation did not penetrate the deck, but the masonry was firmly secured to the surface by means of great ship knees of iron which braced it firmly on all sides and were screwed down to the timbers. It was cased with wood on the sides, or flanks, and the top was completely covered by a large, sloping, battened hatchway.

Removal of the big hatch exposed two or three enormous metal trypots, each having a capacity of more than a barrel. Between catches and trying-out of whales, these pots were kept immaculately clean, polished with soapstone and sand in the best whaling tradition until they shone like silver punch bowls. In the picture from the Nantucket Whaling Museum, the fireboard is shown dropped down exposing the two iron mouths directly underneath the pots, which are fitted with the two heavy iron doors, closed when not in use. The intense heat of the fire was prevented from reaching the deck by a shallow basin filled with water called the "goose-pen" extending under the entire enclosed surface of the works. There was a tunnel in the rear which was kept replenished with water as fast as it evaporated. The smoke escaped from the openings in the back wall as there were no external chimneys.

The ritual of trying-out the blubber followed as set a pattern as applied to the chase and capture of the whale himself. Wood was used in the initial firing of the try-works and the accumulated shavings contributed by the ship's carpenter insured a quick blaze. Once the pieces of blubber were tried out, they shriveled up into scraps or fritters which still contained some oil. These scraps were fished out of the pots and placed in a container to be used for refueling. They produced a clear, fierce flame. The smoke from the try-works was sullen, black and described as evil-smelling as that which rises from a funeral pyre in India. Day and night this ugly smoke curled from the try-works, smudging the sails, blackening the masts and covering the deck with a coat of black grease. When it rained and the vessel had no protection built over the try-works, the process was halted and tons of blubber were put through the hatch below deck.

There are many important steps before the blubber reaches the pots. Blubber is the whale's insulation against cold and pressure, a blanket of fat measuring from 8 to 24 inches in thickness depending upon the species. It is of the consistency of firm, close-grained beef, more elastic and compact and is so tough only the sharpest knife can cut it. The removal of the blubber is the *flensing* process. After it is worked loose by sharp spades, it is cut into what is known as "horse pieces" measuring 3 feet in length, 6 inches in width and 5 inches thick. These pieces are pitched forward to men called "mincers." They cut the pieces on a wooden horse with a large tub beneath to catch the slices which are now prepared for the try-pots.

A very wide knife allows the mincers to cut the blubber into thin slices like bacon which in the language of whaling are known as "books or bibles." The thinner the slices are cut, the better the heat can penetrate every part which speeds the trying-out of the oil. It also increases the quantity produced. The mate continually urges the mincers to slice carefully and thinly by the cry: "Bible leaves! Bible leaves!" The mates attend to the try-works in the process called by landsmen *trying-out* but to whalemen it is plain *boiling* or *stewing*. The officers stand on a platform over the fore-hatch just forward of the try-works; catch the books on a sort of pitchfork and fling them into the big pots from the tub on the port side.

It takes an hour for a kettle full of blubber to be transformed into oil. The *bibles* now crisp, shriveled and of a golden brown color rise to the surface and float. If a drop of water sounds as if it has hit a hot stove, the mate says the oil is properly "cooked." A skimmer, or shallow strainer, skims off the scraps, first letting them drain their oil. After the scraps are removed, the mate takes a copper bailer holding a gallon and ladles the clean, sweet-smelling pure oil into the "cooler" or cooling tank which holds about six barrels of oil. A valve in the bottom is opened and the oil is allowed to run through a long pipe into the storage tanks in the hold.

It takes three days to try-out a large whale. During these times the trying-out watches are six hours each and consist of half of the crew at a time. The work proceeds without stop from day into night and night into day. A whaler must always be ready to receive the next whale, and whales are often taken during the trying-out. The log of the Ship *Elizabeth,* May 18, 1838, in the Pacific tells of some very busy periods. "Began cutting-in at noon, sea nearly calm, had three whales hoisted and three cut-in." Another entry dated July 23rd tells of the starboard boats striking and killing a whale which was gotten alongside by 5 P.M. and secured for the night, while "boiling first whale continued." During this work, whales were again "raised," or seen, at 7 P.M. Two boats were lowered, which were all that could be spared. The chase lasted an hour. The men returned at 8 P.M. "with no whale." Besides having the ship ready to receive more whales, there was another reason for haste. Flensing and cutting-in immediately was economical for much valuable blubber oil exudes through the thin porous skin when the whale is left alongside the ship.

Notwithstanding the delightfully adventurous aspects of whaling, the new hands soon learned that whaling was an untidy business full of very hard, repulsive work. The experienced salts consoled themselves that the work was dirty, but the money was clean. There is no gambling or dishonesty about a whale. No oil. No pay. Everyone shares in greater or lesser degree by his capture. Small consolation this whaleman's philosophy as the scrubbed decks and tidied ship becomes a floating charnel house. A fish is comparatively bloodless, but a whale has tons of red flesh which oozes the scarlet fluid when dissected. The crew is not only oil-soaked but blood spattered. The men wade in blood.

The eternal presence and odor of blubber and boiling oil become unbearable. The whole ship is bathed in oil. The decks are sliding with it. The men are covered with it. The oil is in their hair, on their eyelids, in their ears, under their finger nails, soaked through their shoes and every thread of their clothes. Their food is permeated with it. Their sleeping bunks are drenched with it. Perhaps the only way they endure the drudgery of grease and grime is the recollection of the high excitement of the chase and capture. The thrill is still warm in their veins as they are submerged in an endless cycle of toil. In waking hours, like oil-sodden ghouls, the men are forever hurrying up and down deck hatchways or stand numb with fatigue before the glowing fires.

Having prayed so long for God to send them whales, the blessing has become a curse. And all of their activities center upon the altar of the try-works, where the sacrifice bubbles so furiously. A new watch relieves the old and the routine continues of cutting, mincing, tossing into the seething caldrons. The ship plods forward into the blackness of the night with the blazing ovens filling the dark seas with weird and lurid glows which emblazon her watery path. In heavy seas, the vessel pitches forward or sideways and the scalding oil threatens momentarily to leap out of the imprisoning kettles to engulf the tenders or high priests of the whale cult.

CHAPTER 12

Lengthy Voyages, Scrimshaw and Gams

So be cheery, my lads, let your hearts never fail
While the bold harpooner is striking the whale!
—Nantucket Song.

The trying-out on board ship enabled whalers to continue their incredible voyages for years on end. As early as 1730, crude whaling vessels were equipped to try-out the blubber at sea and to store the oil in barrels. Often the cargo was sent home by packet ships by discharging it at some port such as Valparaiso, Chile, Honolulu, Hawaii, or St. Helena, in the Atlantic Ocean 1200 miles west of Africa. The Bark *Cavalier,* for instance, returned to Salem in 1837 with 75 barrels of sperm oil, having sold 980 barrels of whale-oil at Rio Janeiro. The Ship *Emerald* relieved herself of 100 barrels of sperm at Hobart Town, Tasmania, Australia, when a cruise to the Indian Ocean yielded 2100 barrels of sperm and whale-oil. In its 1840-43 voyage the Ship *Mount Wollaston* sold 1,200 barrels of whale-oil at Rio Janeiro and still returned to Salem with 400 barrels of sperm, 700 barrels of whale-oil and 5,600 pounds of whalebone.

The number of voyages a whaler completed was limited because each embraced so long a stretch of time. In point of voyages made, the *Izette* seems to have been Salem's most active whaler, with six voyages to her credit. The *Emerald* made five voyages, the *Bengal* and the *Reaper* four each and the *Messenger* and *Falcon* three each.

A great deal of time was consumed in reaching the whaling grounds, especially when these were in the Pacific Ocean which meant sailing around Cape Horn. This took two months alone. In addition to the voyage itself to reach a destination, there were weeks spent in cruising about in search of the elusive whales. Sometimes a clue would bring a captain thousands of miles, only to find on reaching the whale-haunted waters, that many vessels were already there hunting and frightening away the schools. During the weeks on board a whaler when there was no whaling activity, a number of things occupied the men in work and play.

During periods of leisure, the crew made spun yarn and *sennit*—braided cord from plaiting rope-yarns together. They ground the heads of their harpoons and lances until they were sharp enough to afford a clean shave. When all the equipment was in readiness for the first spouter, and the insides of the try kettles were polished to a gleaming brilliance, some of the men turned their attention to making knick-knacks out of shells and whale's teeth. It was the Yankee whaleman who created the folk art of *Scrimshaw* during the early part of the eighteenth century.

In Nantucket the carving, etching and inlaying designs on the Sperm's teeth and on the jaw pan bone, were known as *Scrimshant*. The Mystic Marine Museum tells us that the term Scrimshaw comes from an old word *scrimshander* or *scrimshanker* which means an idle, worthless fellow. Gradually the term came to mean the objects made by a sailor during his idle time at sea. Aside from the work of the American Indians, Scrimshaw is the only indigenous American folk art. One of the earliest references to this art by name appears in the log of the Brig *By Chance* under date of May 20, 1826: "All these 24 hours small breezes and thick foggy weather, made no sail. So ends this day. All hands employed scrimshanting."

The only tools used were the jack-knife, small saw, a sharpened harpoon fragment or sail needle or bodkin to etch in the design. The ship's grindstone or file smoothed down the *ivory*, or whale's teeth. To give color and depth to the work the incisions were filled with black, red, and blue India inks. Wood ashes and the palm of the hand brought a high polish which offset the dull colored lines. Almost every museum has on display this lovely art of forgotten whalemen. Free time during a long, lonely voyage translated dreams and aspirations on *scrimps* and scraps into pieces of lasting utility and beauty.

Memories of home, romantic thoughts of sweethearts left behind, memorable incidents of the cruise, the impressive outlines of the old whalers in which they sailed, found expression on chessmen, jewel boxes, cane handles, swifts for reeling yarn, paper cutters, embroidery tools, and the popular jagging wheel for crimping pies and pastry edges. A great favorite of the boys was the busk, or whalebone corset stay on which many a touching sentiment was etched. There was no finer gift for a loved one than a handsome piece of Scrimshaw.

In three centuries New England has created many articles of fine craftsmanship which reflect the region's love of liberty and free enterprise. The numerous commercial pursuits arising out of a land ill-adapted to agriculture have brought to America some of its finest cultural contributions. An outstanding example of this is the whaling industry which was a business, pure and simple. In its actual prosecution whaling had little glamor or esthetic appeal, so little, in fact, that few artists of the time troubled to depict the ships in which men whaled or the processes connected with whaling, including the exciting hunts. Now, as sailing vessels fade into history, the crude whale-oil lamps, the whaling pictures and the scrimshaw which the whaling age produced are eagerly sought by collectors.

There was a social art, too, which was restricted to the whalers. This was the *gam* which Webster's Dictionary now defines as: "To call upon, chat with, and otherwise communicate with, as the officers and crew of another vessel." Another definition by the same authority: "*Gam*: The exchange of calls and social intercourse which passes between the men on whalers and fishing-smacks when they meet at sea and are at leisure."

At the time Melville wrote his *Moby Dick,* the word did not appear in what he called "Noah Webster's Ark." So he suggested that such a word, in wide use by 15,000 Yankees, be defined as: "*Gam.* Noun—A social meeting of two or more whaleships, generally on a cruising-ground when, after exchanging hails, they exchange visits by boats' crews; the two captains remaining, for the time, on board of one ship, and the two chief mates on the other." The origin of the word may be from *gam* meaning a herd of whales, usually playful when undisturbed by danger.

Fishing-smacks may enjoy their gams, but it is the whalers one envisions as past masters of this art. "There was nothing in the merchant marine" Professor Morison holds "corresponding to the friendly 'gams' or visits between whalers at sea; half the officers and crew of each vessel spending several hours, even the whole night, aboard the other." Gamming was an indulgence which owners worried about in the homeport during the long periods of non-communication with their investments at sea. In 1836 Charles R. Tucker, owner, had written a word of caution to his Captain, Charles Starbuck: "Endeavor to avoid those (ships) that wish to spend much time in gamming—as a lone chance is generally best."

Salem's whalers enjoyed gams along with the rest of them. Always welcomed, the meeting with another whaler meant new faces, news of the world, and a fine opportunity to beg needles and thread, also tobacco and pipes. It meant a chance to feast a bit, drink someone else's rum and to sing some sea songs, whaleman's chanteys, or a popular ballad. The log of the *Emerald,* voyage of 1840, gives us a chance to look in on one of these gams which was enjoyed more than ever because of the misery which preceded it. There was a melancholy time on board as whales had been chased and lost and a "stove" boat meant the crew being drenched, almost drowned, and rescued in nick of time. Many good harpoons were lost and lives sacrificed. Additionally, March 17, 1840 saw the whole ship overrun with mice "which did damage to sails in the sail room." March 23rd reported that the *Emerald* of Salem had been out "ten months" and had secured "1,100 barrels of Right and 260 barrels of sperm-oil"—which represented a great deal of back-breaking work. The hard work continued as four days later, the log keeper William Henry Nichols reports "Got whales and started boiling at midnight. Plenty of sharks around this night. We are off North West Cape of New Holland."

Then comes the glad break in the cruise. There was happiness, of course, in "raising" (spotting) whales, in chasing and getting them, but these were periods of wild excitement and the envisioning of nice profits. Long weeks in company with the same men bred boredom and the bottom of the barrels of humor had been scraped and oft found wanting. There were no new stories or jokes and the old ones were worn threadbare. The opportunity to associate with new human beings is at no time more precious than when the companionship of one's intimates has lost its lustre. But all humans being basically the same, the enchantment of novelty is but a case of old wine in new bottles.

By April the 6th the *Emerald* was cruising in company with the *Wade,* and by 3 P.M. spotted another sail ahead. One pleasure at a time. At 5 P.M. everything was in readiness for an exchange of amenities with the *Wade.* "Lowered our Starboard boat and gammed with the *Wade*" says the log book. This continued for two hours, and then the *Emerald*'s officers "spoke the stranger at 7 P.M. and found it the *Clifford Wayne,* Captain Downes. Went on board the *Wayne.* At midnight each boat's crew went on board of their respective ships." From the Barque *Wade,* "we got 12½ pounds of rivets." In large joyous script across the log were three words: "Plenty of Company."

It was good that the *Emerald*'s whalemen had some fun, for five days after the gamming interlude, the log reports "Most of the crew complains of disentery." George Dodge's pamphlet refers also to the dread scurvy. "When a long time on a cruise, it is necessary to give all hands a glass of fire water every morning to keep the scurvy off, which often proves fatal on going ashore after a protracted period at sea. In extreme cases, the smell of the land will prove deadly." In the case of the *Emerald*'s crew, on April 29th the log reports that "Joseph Ravens is swelled considerable with the scurvy, face and feet."

Five days later, relief was in sight. May 4th saw "Captain Dexter on shore while the ship was lying at anchor in Ampanan Bay." The Captain sent off to the empty larders of the *Emerald*'s pantry "three dozen ducks, two very fat hogs and a few sweet potatoes and squashes." Fine weather the next day added to the festivities of the "Thanksgiving day" celebrated "fore and aft" as the *Emerald* lived "very high on fowles and vegetables, fruit, beef, etc." The high life was a little too much for Charles Ingalls who had to be put in irons, but by May 20th "Joseph Ravens was getting better fast" as a result of the fresh vegetables and fruit.

George Dodge tells of "raising" whales in the midst of having a good time at gamming. There is a wild scramble for the men to get into their own lowered boats. "You see some very fine pulling" wrote Dodge "when two boats from different ships are after the same whale."

A distinction which attaches to gamming is the method of transporting the captain to the whaleship which will be his host. It was customary for the captain of a man-of-war, privateer, or slave ship to sit on a cushioned seat in the stern of the rowboat when he was transported away from his vessel. He manipulated a little tiller oar decorated with gay cords or ribbons as an emblem of his position. In a whaleboat, things were far different. The captain was honored in a manner which was both uncomfortable and even hazardous. Within the rules of gamming, a complete boat's crew must leave the ship in order to be prepared to take a whale if it appears. As such a complete crew included the boatsteerer or harpooner, who occupied the place usually afforded to the commander of a vessel, there actually was no place for the skipper to sit. It would, of course, be beneath his dignity to hold down any lesser seat. He was rowed to the visit, therefore, standing erect in the whaleboat, a peculiarity individual to whaling. His prestige now rested on his ability to maintain this upright position.

His own crew watched him as he took off and he was carefully scrutinized by the men on the whaler he approached. As the captain straddled the whaleboat to keep his balance, he had to appear nonchalant to a painful degree. The casual air required was achieved by hiding his hands in the secretive recesses of his pocket trousers. This bit of whaling etiquette was no easy assignment for the commanding officer. In his rear, the boatsteerer's 28-foot steering oar hit him every now and again in the small of his back. In front, the after-oar maintained the rhythm of tapping his knees at regular intervals. He had also to hold his own against any violent pitch of the boat, until at last the rowers brought him safely to the side of his host's vessel. In a sudden squall there have been instances, Melville testifies, in which the Skipper seized hold of the nearest oarsman's hair and held on there like grim death.

And, in the event the captain's lady was not content to remain meekly at home for some four years with the dubious status of neither wife, maid nor widow—there was the problem of transporting her. The wife of the captain was not required to stand up in the whaleboat, and a "gamming chair" was always provided to hoist her on board. Be sure to look at one of these "gamming chairs" at the Nantucket Whaling Museum. There is also one on the deck of the *Morgan* at Mystic.

It was not too unusual for a whaling commander to bring his wife along. Some ships maintained high enough standards to make this feasible. When the famous *Charles W. Morgan* of New Bedford sailed on her maiden voyage in 1841 for a cruise of three years and four months, Captain Thomas H. Norton was accompanied by his young wife. It became a tradition on the *Charles W. Morgan* for the commander to be thus accompanied up to the last working days of the old whaler. In 1913 we find Mrs. Charles S. Church with her captain husband on a whaling cruise in the *Morgan* which was Mrs. Church's second voyage on this vessel.

It is not so difficult to understand when we recall that pretty girls of Nantucket, New Bedford and other whaling ports would not give a boy a tumble unless he had killed a whale. When he went a-courting, a lucky suitor wore in the buttonhole of his Sunday suit the little block or button of the boatsteerer and harpooner. And when they finally were married, and the bridegroom happened to be a whaling skipper, as lief as not, the bride would accompany him on his three or four year odyssey. Several years ago, George A. Grant, a curator of the Nantucket Whaling Museum, wrote: "All of my mother's children were born in the Pacific Ocean at different places. Brother Charley at Pitcairn Island in 1850." Mr. Grant himself was born in Samoa in 1856 during a whaling cruise.

Salem women sometimes accompanied their captain husbands on trading quests to exotic ports of the world. Within the boundaries of some eighty whaling voyages over a half a century of time, there were women who went whaling out of Salem. Their presence does not appear in the sea journals examined but the untold story is there, hidden away in old letters and notebooks which some day may come to light.

In his remarkable book *Whaling Wives,* Henry Beetle Hough finished out the story undertaken by Emma Mayhew Whiting from unpublished letters and diaries of the women of Martha's Vineyard who went whaling with their husbands. He tells the story of their strange and often alarming life as they rode out storms, met up with cannibals, endured the trials of mutiny on the high seas with no more fuss than baking a blueberry pie. Where there are women, there are children. Laura Jernegan with her mother and small brother sailed the Pacific for years on her father's ship and kept a little diary. After a tremendous catch of whales with the men furiously boiling blubber on deck, Laura sat down calmly and wrote a friend: "We had baked potatoes for supper and biscuits. Would you like to hear some news well I don't know any."

Mr. Hough covers the voyages of thirty-six wives who accompanied their captain husbands on whaling voyages extending from 1834 to 1889. The book even ferrets out the very young of the female gender. For instance when the Bark *Morning Star* arrived at New Bedford on May 15, 1857 from the Pacific, the marine news columns carried the item: "Passenger, Miss Cleveland." Miss Cleveland was four years and two months old and she had made the voyage around the Horn without companion of her age or sex aboard the whaler.

Fortunately for interested posterity, the ladies were addicted to diaries and letter writing. Susan Folger Fisher's lengthy letters from her bleak address "Pacific Ocean" are enchanting. Here is a taste of one:

June 21. Did you ever think what a deceitful creature Dame Fortune was? Today she holds the cup of prosperity to our lips, tomorrow it is dashed in pieces. But I have been deceived often enough not to trust her too much. But where the organ of hope is large we are apt to build large castles. Really, I did think we had struck into a vein of luck, for in sixteen days we had taken something over 550 bbls. of oil, and in the time we had two blows and some fog. But we were among whales, which had a bright look of our making up our last bad season's work.

But the *Cowper* is in trouble now. She has betrayed her trust, but not willingly; however I shall be fearful of her the remainder of the voyage. Last Monday morning at three o'clock she struck a cake of ice and stove in three planks on her bow. (The *Cowper* out of Martha's Vineyard was in the Okhotsk Sea) We were not in the ice at the time, but running along the edge of it with scarcely any headway on the ship . . . The piece we struck seemed to be by itself coming very quickly towards us, and the man did not get the wheel up soon enough to escape it. The shock was dreadful.

N. was already up and was on deck in a moment. He and the mate both looked over the bow and thought the ship had received no injury. It being 4½ feet under water it could not be seen on the outside. They were boiling at the time and had about 250 bbls of oil on deck; so N. thought he would commence stowing down . . . When they went down they heard the water running in like a brook. The mate ran aft and told N. the ship was leaking badly. He examined, and found nearly five feet of water in the hold; rigged both pumps and commenced breaking her out forward. To tip her, however, they freed her first and stopped the force of the water, as much as they could, on the inside. By 7 o'clock it breezed up and we were in a sad situation.

N. set his signal for assistance and a ship near, soon came to us with all four boats. They thought by our appearance, we were sinking. We had now about sixty men on board and they *worked* I assure you. About 5 o'clock P.M. had the ship tipped as much as possible so that the leak was brought just on the surface of the water, when in a short time they got it temporarily mended by nailing on tarred canvas and boat boards. It then took until the next noon (yesterday) to get the ship in trim to sail any, but now we are on our way to a small port by the name of Ayan, to repair. N. thinks it about three days sail from where we are.

In all probability it will be a hindrance of about three weeks. Is it not too bad to lose so much time in the midst of the season. But we should not complain, for it might have been much worse. The day before our accident we spoke the Barque *Shepherdess,* Capt. Waterhouse. He and his wife came on board and took tea with us. She was an entire stranger to me, but we had quite a good gam together. She was only eight months from home but was heartily sick of it. I *comforted* her by telling her that she had seen only the best part of whaling, yet. We have been in sight of three ships that have ladies on board, one was an acquaintance but we were both chasing whales and could not stop to speak.

July 4th. How are you spending this great and glorious day? All I can do towards celebrating it, is to have a chicken pie for dinner which I superintended the making of, myself. Well the *Cowper* is repaired in good shape and we are on the move again. We left Ayan yesterday, where we staid five days, during which time I was on shore with the Russians. N. came on shore every day but was on board the ship most of the time. I met with a hearty welcome among them and found them very pleasant people. They wanted to do more than everything for me, but I could not understand their language and only two persons in the place could speak English. One was the gentleman at whose house I stopped.

The settlement is between two mountains and has about 30 houses and 200 inhabitants, it is a company sent there by the Russian Emperor—there are six families and four or five naval officers that call themselves the nobility, the rest are the laborers. The nobility are intelligent and very polite, their dress is of the best material, their cloaks of nice broadcloth, their winter ones lined throughout with beautiful fur, their summer ones with silk, wadded and quilted, the style the same as in New Bedford last year, as also their bonnets. Some of their customs amused me very much, they always throw a nice shawl over their shoulders to receive visitors . . . we dined at the Governor's and everything was carried on with the greatest formality; we had five different dishes, but neither were suited to my taste. . . ."

The *Cowper,* Captain Nehemiah C. Fisher of Edgartown, sailed from New Bedford in the fall of 1851. Destination—the

North Pacific. The wife of the captain accompanied him and the 391-ton vessel became her home for three years. Susan Folger was from Nantucket and whaling was in her blood. They were in the twentieth year of their marriage when they went together to the Pacific and to the Okhotsk Sea, off Siberia. Susan took the fogs, ice and chilling cold in her stride. She knew the voyage would include the balmy and pleasant sojourn in the South Pacific and it was worth waiting for. From the North Pacific she wrote a cheerful letter to her cousins: "When the weather is good, I go on the decks and watch the boats until I am cold and tired, then come down into my snug little cabin to warm and rest myself. I am growing weary of this long voyage, one season more than we expected, but as everything has an end, I suppose this will . . . The steward has just struck eight bells and called the watch . . . supper is ready."

CHAPTER

13

A Whaleman's Lay From an Agent's Daybook

"A smooth sea never made a skillful mariner . . . prosperity deceives . . . adversity instructs."
Richard E. Burton (1861-1940)

The delightful gams were held forth by shipping agents as one of the inducements to would-be whalemen. In reality, the gams were few and far between. The hazard of the hunt was great. The work during trying-out was almost punitive, and the long period of despair during which there was no whaling activity was the penance which tried whalemen's souls. As like as not, the profits for the effort expended and the deprivation endured would be meager. On the other hand, there were cases of the bonanza voyages about which every whaleman dreamed.

The account book of James Cheever, Agent, Salem, Massachusetts, reveals some very low returns, not only to the crew but to the officers as well. On the other hand we see cheerful four figures. This day book was given to the Essex Institute in 1903 by Edward E. Cheever. Page 1 contains disbursements of the Ship *Sapphire* for a whaling voyage paid by James W. Cheever in 1836. There are three full pages of disbursements of the Ship *Derby,* 1837. Many entries appear on selling whale-oil by the cask, mentioning a price of 31¢ and 32¢ a gallon by the cask, and sperm oil at 93¢ a gallon. Some examples:

Caleb Smith Dr. to Barque *Eliza*	
123 casks of Whale-Oil gauging 15,810½ gallons @ 31¢, May 18.	$4,901.25
George Lambert to Barque *Eliza,* Dr.	
To 9 casks, 1584 gallons Sperm oil @ 93¢, Oct. 21.	1,473.12

Caleb Smith to Barque *Eliza*
To 45 casks, 6,153 gallons Sperm oil @ 92¢, Nov. 8 5,660.76

Here are the disbursements of the Barque *Eliza* in payment of officers and crew by James W. Cheever, Agent for the voyage which began November 21, 1838 and ended May 9, 1841:

1841, May	10.	Cash paid entering at Custom House	$ 3.36
"	12.	" " Capt. Radcliff draft for provisions to owners of Ship *Bowditch* and interest	78.74
"	"	" " W. H. Osborn, expenses to Boston 4 times a/c *Eliza*.	4.75
"	15	My expenses to Boston	1.00
		Postage to date	.95
"	18	To Officers and crew as follows:	
		Capt. W. C. Radcliff on a/c	60.00
		William Bennett, 1st Officer, balance of his acct.	789.04
		Jared Wade, 2nd Officer, ditto	473.27
		Lewis Laurence, Cooper "	170.92
		Edward Holmes, "	97.69
		John Malzard "	106.33
		Benj. Berry, "	29.98
		Wm. Berry "	67.61
		Wm. Mills "	44.58
		Edward Gretney "	44.80
		John Adams "	70.76
		Thomas Sluman "	56.36
		Hiram Hand "	26.59
		Augustus Robinson "	25.62
		John Oliver "	5.39
		George Robinson "	84.85
		John Woodworth "	58.28
		Antonis Losell "	96.37
		Charles Robinson "	24.54
		Paid expenses discharging, cleaning & C. P. a/c	92.50
		" John Williams, Cooper	9.52
		" J. R. Call	6.75
		" Ben. Peach. Ship keeper	10.41
		" wharfage and dockage & storage	51.33
		" J. Sanborn, carting ore to shed	22.00
June	2,	paid W. C. Radcliff on a/c his voyage	60.00
"	15	" " " " " " "	700.00
July	29	" " " " " " "	500.00
"	29	" Nathaniel Cleaves, gauging oil	24.50
"	"	Pilotage inward to Salem	14.60
Aug.	11.	Cash paid W. C. Radcliff order in favor J. Melzard	22.40
Sept.	23.	Paid premium note C. S. Company	813.96
"	"	" " " O. Insurance Co.	813.13
Oct.	21	" " " Ocean Insurance Co.	61.00
Nov.	5	Paid proportion of expenses to Portsmouth & New Bedford a/c oil	7.00
		Paid rent of shed per bill	26.89
			$5,657.77

" Capt. Radcliff balance of his lay viz 1/16 of $21,625.75		1,351.61
Less advanced above		– 1,342.40
		$ 9.21
	$5,691.92	
Adv. amount of postage	+ 3.55	
	$5,695.47 Disbursements	

When George Dodge went whaling from Nantucket in 1831, each of the crew was to receive the value of one barrel of oil out of every 180 barrels secured. The captain was to get one barrel out of every 14 secured, the mate one out of 40 barrels and the other officers in relation to their rank. The owners provided the whaling outfit and charged 25% interest on the money, so says Dodge, "A large slice of the voyage was gone before being earned." After 44 months at sea, or four years lacking four months, George Dodge received the handsome sum of $125.00. All his debts had been paid out of his earnings. He bought some new clothes in Boston and took the old Eastern stagecoach home to Salem to tell family and friends of his whaling adventures.

The pay on a Salem whaler tended to be less than that prevailing in larger whaling ports. Able whalemen found their way to Nantucket, New Bedford, Fairhaven, New London, etc. However, a goodly percentage of whalemen from these famous whaleports are found among the crew lists of Salem whalers. Salem whaling merchants and agents experienced the usual difficulty in shipping enough men to share the work.

The earnings of the *Emerald* which sailed out of Salem Harbor in 1836 represent a poor voyage. The *Emerald* shipped 24 men and was absent for 22 months. At the end of the voyage only six men, including the captain, received over $100.00. Nine of the crew owed money to the owners and had nothing but their debts and harrowing experiences to show for the two-year stretch.

A more successful outcome appears in the records of the

Ship *Elizabeth* which left Salem under Captain Hedge on a whaling voyage in 1836. This voyage happens to be one of the most successful ever launched by Salem capital. The rewarding cargo included 2400 barrels of sperm oil worth $72,000. At the end of three years and six months, Henry Cotton, officer, received $2,264.21. Abner Lake, officer, received $1,743.28, and the other men received sums ranging down to $55.14. Nobody owed the owners anything and there were only four men who deserted ship.

Responsible seamen could usually find employment on cargo carriers at guaranteed wages. It is easier to appreciate the returns on a whaler by the lay or share system by drawing a comparison between whaling pay and earnings in other types of maritime work. The merchant marine at this period was paying high enough wages, but in whaling there was the ever present possibility of sharing in a very successful voyage, which might even include a catch of ambergris, worth its weight in gold. When William Nichols returned to Salem from his whaling voyage as a cooper on board the *Emerald,* he soon went to sea again. His subsequent earnings on a commercial carrier were recorded in the back of the *Emerald*'s log, which Nichols had kept during his whaling voyage. From these notes it appears that the commercial voyages were more remunerative than had been his whaling work. Let us see.

Nichols left Boston on September 17, 1840 for a nine months' voyage to Cienfuegos, Cuba, in the Brig *Eliza Burgess.* He worked under contract as a cooper for Messrs. Benjamin Burgess & Sons at $40 per month with board and passage provided. For six months' work, Nichols received $240 plus $95 for extra work. For acting as 2nd mate for the *Eliza Burgess* on the return voyage, Nichols earned an additional $20 a month. As the voyage did not last a month but only 19 days, Nichols received $12.60, "after a rough passage to Boston." His notations close with this statement: "Nine months' work plus extra earned as 2nd mate, netted $532.21, deducting passage out and passports of $40, my net was $492.21."

It is significant that Nichols was able to earn at an average of $54 a month during nine months in a commercial voyage devoid of any unusual risks or athletics. Comparing this net profit to that of Henry Cotton the highest paid man on the crew of the *Elizabeth* in her voyage of 1836 (rated one of the "most successful" out of Salem), we find that Henry Cotton's earnings of $2,264.21 for a period of 36 months averages out to about $53.00 a month, or a trifle less than that of William Nichols' voyage on the *Eliza Burgess.*

A poor voyage meant many desertions. Of the 34 men who shipped with the *Sapphire* in 1836, 17 men deserted, or 50%, which was 25% above the average. Some names picked at random from Mr. Cheever's accounts show a wide range of compensation. The largest share was—outside of the Captain—received by John Crowell, Mate. At 1/27ths the value of one barrel out of every 27, his earnings came to $881.30. Stephen Briggs received 1/60th or $615.07. Josiah Breed 1/70th or $342.64. John Snelling received 1/65th or $31.20. Charles Gray received 1/150th or $3.95. Charles Williams *owed* the owners $3.83. John Anton *owed* the owners $34.28.

These debts were for the purchase at high prices of a rather good grade of boots, clothes, tobacco and rum from the ship's "slop chest." Not all captains were heartless villains, but George Dodge's skipper, Captain Chadwick who was sixty, had seen enough of the world's nakedness, to make him callous to the niceties of life. "If a sailor lost a hat overboard," Dodge lamented "if it was ever so calm, he would not lower a boat to pick it up, but would laugh saying 'I have plenty to sell.' Some of the crew had no other way of getting money when they were in port but to draw clothes from the captain and sell them on shore for half of what they paid for them."

The logs carry expressions of despair in the long drifting and waiting for whales, and over the affliction of bad weather. "It is lone-enough for anybody" wrote the mate of the Ship *Elizabeth* under November 23, 1845 "but we will hope for the best and endeavor to make the best of things by submitting to those things which cannot be controlled and that the time will hasten on which will return us to our loved ones again is the fondest hope of my heart." The next day brought "Squally, heavy swell of seas enough to make a pirate of a saint. Thus we go and have gone for these 34 months, and God in infinite mercy send the end of this cruise and restore us to our port of destiny, but there are a few that say I make a long story of nothing."

The complaint of the Ship *Bengal*'s log keeper was the lack of wind, whales and food. The sails drooped. The ship stood as still as the *Ancient Mariner's* "painted ship on a painted ocean." Under April 23, 1840, we find this entry of hopelessness: "Calm and very like to remain so for a month. For a month now we have had hard times. There is a long road ahead, very little provisions to eat. I do not know what will become of us. They say it is all the old man's fault. I suppose it is. We must bear it. Lord of the sea, send us a breeze of wind, I pray thee sincerely lest we all perish—wicked sinners as we are. O my home, my own native shore! When shall I see thee more? In good time, I hope. Trust for the rest to kind Providence with whom all things are possible and without His help we can do nothing."

Another lonesome logkeeper was found pouring out his heart in the late night watch, on board the *Izette,* Bay of Whales, the Ross Sea in the Antarctic (1845): "I want to hear from home very much. This whaling business is not all that it is cracked up to be, let me tell you, and is not half so profitable as some have been foolish enough to suppose. I want to place my foot once more on the sod and see some new faces and hear some new stories for it is an act of charity now to listen to the old ones. I often think it wonderful how a man can exile himself so long from friends and from home and knock about the ocean for so greasy an object. It is a life of trouble and toil at the best and I suppose I must have my share as well as the rest of mankind."

And why *did* a "man exile himself" for so many years from civilized life? That remains the whaleman's secret. Very few who wandered to the far corners of the earth found the pot of gold at the end of the rainbow. Many were called to the service of the sea, while few were chosen for its rewards. What price the "glorious" life of a whaleman?

The Salem logs have the usual amount of entries about men being flogged while tied to the rigging, put into irons after denying duty or for becoming intoxicated on shore leave, or overstaying a liberty. I have not seen any evidence of hazing which was prevalent especially on whaling vessels. A certain ceremony took place on some whalers crossing "the line" or equator, and no doubt Salem whalers went along with the times. It was the custom on board some of the old-time whalers to subject to an ordeal all those men who had not previously crossed the equator. This hazing was facetiously called an "introduction to Father Neptune." Each candidate was in turn strapped in the "gamming chair" and "dunked" in the ocean until the bubbles of air stopped rising, then he was pulled up. Sometimes the "greenies" were further chastised by being "keelhauled." This meant they were pulled completely under the vessel from one side to the other. As if this were not sufficient, a coating of tar and lard was applied to the victim's face and scraped off with a piece of hoop iron. The survivors were considered duly initiated into the brotherhood of the sea and roundly congratulated.

A federal statute of 1835 sought improvement of maritime conditions by prescribing punishment of officers "who from malice, hatred or revenge shall beat, wound or imprison a member of his crew or inflict any cruel or unusual punishment." As each captain was a law unto himself, circumvention of the law prevented any marked improvement on some vessels. Often, when making a port, liberty was denied and to prevent desertions the ship was anchored off-shore. The captain or his mates rowed into town for provisions, recreation and trade. Intolerable conditions at sea were described in Richard Henry Dana's book *Two Years Before the Mast.* Published in 1840, this authentic account did more than any statute toward the correction of maritime evils.

While some writers sublimated and fictionalized the Pacific, others wrote hard-hitting books born out of personal experience. In addition to Dana's towering classic of maritime brutality, there were writers who pictured the immense world of the Pacific which dwarfs all else into fantasy, transforming a man into someone else. Whatever he is, the Pacific makes him more so. Captain William Bligh, the inflexible, became more rigid and ultimately, as we know, even monstrous.

Frank Bullen's *The Cruise of the Cachelot* gets to the heart of the whaling business. Victor Slocum tells of the *Life and Voyages of Capt. Joshua Slocum,* his mariner father, in which he includes maps and drawings which preserve forever the true story of the first man to sail around the world—alone—in a 36-foot sloop called the *Spray*. The list is quite long. Not to be omitted is *Whale Hunt* by Nelson Cole Haley, the narrative of a voyage by the harpooner in the Ship *Charles W. Morgan,* 1849 to 1853, as taken from the original ship's logs. An early vintage is *Narratives of Voyages and Travels* in which an American sea captain describes his experiences in three voyages around the world, published in 1817. The author is Amasa Delano who was born in Duxbury, Massachusetts in 1763, soldiered in the Revolution and afterwards became a privateersman.

There are many tales of the Pacific. Even those woven of the truth defy credibility. Essentially all present the premise of man against the sea, man against man, adventures of sailors tossed upon the breast of the Pacific in small boats which put literature on another scale. With a little imagination, the old sea journals are revealing.

Let us open the old log of the Ship *Sapphire* which went whaling in the Pacific Ocean, and note the struggle of the crew against the captain, and the captain against conditions. On April 28, 1836 at 10 P.M. "Paul Sweet, James Lewis and William Hall for mutinous conduct were taken out of the ship and carried on board the U.S. Frigate *Constitution.*" Putting the men on board *Old Ironsides* placed them under the surveillance of the United States Government. Four days later "the three men are taken aboard from the *Constitution* after promising good behaviour."

The next day, May 3, the "steward and the third mate returned to the ship after being absent 48 hours—likewise the cook after being absent 48 hours." Paul Sweet obtained a new friend, John Adams, and both deserted. The log tells how they were found on board the Ship *Heroine* and were taken and put in the calaboose, while the boat in which they had deserted was found drifting out of the harbor.

On the *Sapphire*'s voyage of 1842, the officers had occasion to rescue men who had deserted from another ship. April 25: "Steering southwest fell in with a whaleboat with six men in it, deserters from the Ship *Martha* of Fairhaven. They had been in the boat for nine days and were destitute of water and provisions. Supplied them with 50 gallons of water, 2 pounds of bread, 50 pounds of beef and pork. Gave them a compass and told them how to steer."

We find William Hall (mentioned as a deserter in the *Sapphire*'s voyage of 1836) a malcontent on the present voyage of this whaler. On July 16, 1842, he is among "the group of eleven men who came aft and inquired of the captain if he intended to ship more men and he told them the complement was now 24 and he considered that number sufficient to carry the ship home, but if the consul did not consider it enough, he would ship more and ordered them forward and went on shore."

"The next day the captain came on board at 4 P.M. called the crew, and told them he should not ship any more men, that the Consul considered the complement on board now sufficient. At 5 A.M. he called on all hands to get under way. The most of them refused to take up the anchor until the captain shipped more men. The captain came forward and called on every man and asked them separately if they denied duty. The following men declared that they did: William Hall, Edward Ishamahiel, John Cambel, Henry Flint, Charles Rideout, Manuel Brown, Joseph Sabino, Joshua Chase and Timothy Fisher. The captain then went on shore and in about an hour sent for Hall and Ishamahiel with their things and at 8:00 A.M. came on board, and ordered that all that had denied duty to be tied to the rigging. They came in a threatening manner with clenched fists, threatening to strike us. John Cambel calling on them to come on and let us have it, but we succeeded in tying them. The captain then gave them 12 lashes each and then asked them if they would go to their duty which they promised to do. The captain then went on shore—and shipped two more hands." The strategy is obvious. The captain kept his word "to ship two more men." They apparently were Hall and Ishamahiel whom he had ordered off the ship that morning.

Shore-leave usually meant breaking the rules. Beside spending all the money they could get their hands on, the men imbibed too freely and forgot to return to the ship. *Sept. 13:* "The boat came from the shore with the liberty men in it. Stephen Lambrose, excepted, nearly all were intoxicated including the 2nd and 4th officers. Mr. Shepherd was badly bruised. At 8 A.M. Stephen Lambrose came on board having been 14 hours over his liberty." One week later the *Sapphire* was still in port with the customary results. *Sept. 20.* "Only a part of the liberty men came on board. At 8 A.M. the captain went on shore to get the men on board. The cook Francis Jones resisted his order and the captain had to call assistance to tie him and take him by force on board. At 9 A.M. he was put in irons on board the ship. The rest—John Fish, Edward Ishamahiel, Manual Brown, Abner Burman and James Lewis—came on board at the same time after being absent 14 hours over their liberty, mostly intoxicated. Anton Lemir is still absent."

CHAPTER

14

Romance in the South Seas

"Romance has been elegantly defined as the offspring of fiction and love.'

Benjamin Disraeli (1804-81)

There are many entries in the logs of shore-leaves ending in overstaying liberty, returning intoxicated or not returning at all. The reluctance of the men to report for duty, once they had enjoyed leave on a tropical island is not surprising. It was natural to become irked with the smell of dead whales, the tortuous cutting-ins and trying-outs at sea, and the eternal mewing of the sea birds. Long poring over the yellowed pages of the old whaling journals convinces me that some interesting entries were never made in the record which had to be examined by the owner or agent at the close of the voyage. Captain Neal of the *Reaper* was so outraged by the actions of his crew, he did not refrain from writing it into the journal. But this is an exception.

There are glimpses here and there of an angle of the industry which has been carefully overlooked in whaling literature. Granted there were many reasons why a man overstayed his liberty and even went so far as desertion, still not all of the reasons pertained to the unpleasantness of living on a whaleship. This was to be expected and generally accepted. For some men there was the memory of an irritating family obligation at home, unpaid debts, or the uncertainty of employment on land once the cruise was ended. These were all negative reasons. What can be said about the sheer enchantment of the Pacific Islands as a powerful inducement to desertion?

Here we find a light and fragrant side of whaling which New England has discreetly kept in the dark. For some Salem to Samoa, Hawaii, Tahiti rapidly became a one-way trip. Many shipwrecked sailors, the Robinson Crusoes or "lone men" of whaling, soon consoled themselves by marrying native women, often of high rank. Several even ruled over an island the remainder of their lives. The oft-given example of David Whippey was a boyhood dream of New England boys. David sailed from Nantucket in 1819, landed on Mba in the Fijis, and shortly thereafter met a native princess. They fell in love and David Whippey remained to become a chief. It was in the Marquesas Islands—a familiar whaling ground for Salem whalers—that Herman Melville found the living inspiration for his book *Typee*. Sailing out of New Bedford in 1842 on board the whaler *Acushnet,* he formed a romantic attachment in the Marquesas with Fayaway, a beautiful native girl.

New England whalers had reached the South Pacific in the eighteenth century. The first of the lumbering whaling vessels to arrive was the *Beaver* from Nantucket in 1791. A wonderful welcome awaited captain and crew of the *Beaver* after she had safely weathered the rounding of Cape Horn, so stormy that even in summer months the gales ripped viciously at the sails, snow and ice encrusted the ship's rigging. But after the ordeal was over, the *Beaver* stumbled into Paradise.

Here the senses are assailed from every side. First and foremost, there are the depressing distances and awesome space of the Pacific, which covers a third of the earth's surface, more than all of the continents combined. Largest ocean in the world, its reputation was well known. Dreaded for its vastness, the Pacific meant the crew would suffer death by scurvy, heat, assorted afflictions and the calms which arrested all progress of a vessel. Nevertheless nature balances the bad with good.

Usually calm, even to monotony, winds come of 75 miles an hour which produce staggering waves. At such times violence makes a nightmare of gigantic combers which can lift and drop a ship, sometimes in fearful plunges of sixty and eighty feet. In fact, the highest wave on record, 112 feet to its crest, occurred in the Pacific Ocean. In a rivalry of the elements, wind and water war with each other causing to rise from the surface a dense haze resembling smoke.

Here man is dwarfed by nature; in the massive emptiness, he is no longer in control and he realizes it. Everywhere there is this strange awareness, this shared sense of awe and fear at the "very brutal cosmic force and size of the place." There are at times earthquakes which bring frightening tidal waves. Typhoons of tremendous ferocity sweep away small islands, uproot coconut trees, shooting them skyward like so many shafts from an arrow. Nevertheless, the Pacific Ocean, as its name implies, is normally peaceful, and arriving from around the Horn or down from the frozen Northern Pacific with its wilderness of cold, ice, seals and Eskimos, the whalers moved gratefully into the southern whaling grounds.

Usually, the waves, clouds, rain squalls, and the currents are predictable. Even a few days south of Hawaii, the northeast trades prevail with gentle breezes, and then south into the equatorial calms where the sun burns closer, monsoon rains come, vegetation is lush and the colors brilliant. At night the water is radiant with its own kind of beauty. In his book *Voyage of H.M.S. Beagle,* Charles Darwin described the peculiar phosphorescence emanating from the sea. "Every part of the surface which during the day is seen as foam, now glowed with a pale light. The vessel drove before her bows two billows of liquid phosphorus and in her wake she was followed by a milky train."

The whalemen's first impression was the freedom from responsibility enjoyed by the islanders. Women clothed themselves in pandanus cloth sarongs, men in nothing at all, short skirts (*sulus*) or just a penis wrapper. The sinewy-muscled, sun-tanned New England whalemen, with a wholesome attraction of their own, were well received on their arrival in the

Eden of the Sunny Isles. Smiling tawny-skinned mermaids swam in numbers to the ships, a prettier sight, to be sure, than schools of whales. The athletic girls pulled themselves up to the decks by the anchor ropes and the old whalers rang with merriment. A warm welcome also awaited whaling adventurers who leaped from the whaleboats on to the beaches. Flower-decked, dark-eyed and barefooted girls threw fluffy *leis* around the necks of the newcomers.

The world has long known of the feminine charms of the islanders. The women of Tahiti are perhaps the loveliest in the South Seas. Of all the archipelagos in the Pacific, the island of Tahiti, in particular, has been the dream island of the Western World, with a special power over the minds and desires of men. Without doubt it was the awaited Paradise of the ship-jumping sailor and whaleman who answered the siren call of its child-like people and the magnetic appeal of their habitat. Granted that Captain Bligh was a cruel master, some writers say the call of the island is what really caused the mutiny on the *Bounty*.

All white men were intruders in the Pacific. It attracted mariner, adventurer, explorer, profiteer and writer. The scriveners described in their own way the relations of the intruders with each other (under the influence of the islands) and their relations with the natives. They told how the lives of the outlanders touched the people who really belonged there, and, in return, the effect upon the white visitors. What basically were the motives of those who rounded Cape Horn and steered into the Pacific? Vincent McHugh, a modern writer, puts it concretely: One part discovery, conquest and profit (not very much) one part the eternal dream of bucolic ease (still there, if you can find it, and if you can stand it when you get it), plus two parts a vision of happy sexuality. Of these motives, he holds, the third had decidedly the most attraction for most men.

Bougainville in France, Captain Samuel Wallis (who went

to Tahiti in *The Dolphin* in 1767) and Captain Cook in England, Herman Melville in America, each were responsible for creating the Pacific Myth which merged later into the Great Pacific Myth. This centered around some sexual encounters of barbarians from Europe and America with some island peoples. In the beginning, it seems that the crew of *The Dolphin* set a pattern of the world's oldest profession. In their ignorance of island ways, the sailors thought they had discovered at Tahiti the trading price of a girl—one iron nail. This led to the danger of the ship falling apart. The nails were valuable to the Tahitians but the girls themselves had no real use for the nails. It seemed to be expected of them and so made for some coy bargaining. From then on, visitors had the notion that Polynesian women were willing at low prices.

While Captain Cook and other intruders eventually learned that Polynesians had sexual ethics which could be as complex as any other people's, the impression was left upon Tahitian women that European and American men lived in countries where there were few women and men had to bargain for their favors. Therefore, they concluded the distaffers of these very travelers must all be prostitutes. It was the natural misunderstanding and misinterpretation of one code for another.

As for native men, their opportunity of any social intercourse with white women was little. White women were very scarce in the Pacific and regarded with much curiosity and awe. Some whaling captains brought their wives and children with them and when the missionaries came they brought their families. There were some cases of female adventurers in the Pacific. We know of Barbara Thomson, a captive of the natives, whose story is much like that of Mary Jemison's life among the American Indians in New York State. Then there was Mlle. Jeanne Baret (or Baré), a true adventuress whose feats, unfortunately, have been little known because of literary neglect. Jeanne Baré, nevertheless, was the first woman to circumnavigate the globe. Being an orphan, with no family ties, and having lost her patrimony in a lawsuit in France, she ran away to sea. This was no easy undertaking for a girl, so like most of her early sister-adventurers, Jeanne did it by impersonating a man, shipping out as Monsieur Baré aboard Monsieur de Bougainville's frigate. Her duty was to serve the naturalist Monsieur Philibert. When her secret was finally discovered after the ship put in at Tahiti, it was quite embarrassing to the old Commerçon. But the Tahitians were enormously excited because they had never before seen a European woman.

It is good that a number of mariners felt a compulsion to record what they experienced. Some of the writers had worthwhile adventures of their own. Others went to the islands and simply recreated the fantastic experiences of others. Charles Nordhoff and James Norman Hall lived on Tahiti and co-authored book after book, three on the *Bounty* theme: *Mutiny on the Bounty, Mutiny* and *Men Against the Sea.* The latter volume describes Captain Bligh's voyage in an open boat for 3,600 miles and gives a full account of Pitcairn Island where the mutineers fled from Tahiti.

From the far-off days when whalemen deserted for an islander's life until the present, Tahiti has been synonymous with Romance and Escape from Reality. The pattern was set by Herman Melville, Robert Louis Stevenson, Pierre Loti, Rupert Brooke, Jack London, whose books spread the dream. Lord Byron, who considered himself a connoisseur of love, also wrote a book entitled *The Island,* in which he made it quite clear that natural passions could find expression in the islands. Diderot of France wrote that the natural life was superior to anything civilization could offer.

The debunkers say these early enthusiastic literati were weavers of words, that actually Melville knew more about whales than about women, that Lord Byron never had been to the South Seas and that Diderot had transformed the journals of Bougainville into fiction. To be fair about it, what really happens is that the foreigner, in distance alone, has arrived on the other side of the world. He comes armed with his dream and for him it transforms the reality of the place.

Even those who do not come are fired by their imaginations into glowing prose. With its mesmeric emptiness the Pacific intensifies the emptiness within the white man. He knows instinctively he does not belong in this island world. For some, therefore, union with a native woman is temporarily reassuring. It is the old story. By steeping the senses in forgetfulness, the intensity of momentary experience is relief to boredom and the overpowering languor which space imparts.

Young men from New England came from families impregnated with Calvinism and Puritanism; they themselves were well inculcated in the "proper" ideas about the great mystery of life. To many Westerners, sex without benefit of clergy, is sin. It means to involve one's future, to become obligated in a dramatic act of consequence. On the other hand, in the world of the Pacific, the realization comes quickly that here this does not seem to be true. For one thing, the women are different. Their secret is to impart the impression of being virginal and yet wanton, which is exactly their purpose. For one thing, to the Polynesian, love is a simple thing, a pleasurable and satisfying exercise, akin to swimming, eating, sleeping, an isolated delight without serious consequence, yet performed discreetly, without any sense of guilt, no web, no need for the lover to murmur promises of eternal devotion.

Yet, with its tremendous potential for lovemaking, something is amiss in the Pacific for the white man. It is the attitude of native women which takes the excitement out of love affairs with them. It is somewhat disconcerting to a man's ego that his rapt, simple, dignified and enticing partner can be just that to her next lover. This is against the whole sex code of the Westerner who has been taught that union of sense is lust, the union of souls is love, a premise devised by self-made arbiters of time immemorial. In one sense, it is true that there is little genuine thrill of conquest. The islander surrenders only in body and emotions. In some mysterious way her soul remains aloof. She does not and cannot see the male as the only man for her, nor does she consider that sexual union makes her his.

While incidental to the girl once it was over, the remembrance of an exotic union with a lovely native often lingered to torture the mind and the imagination of the white man with exquisite memories coupled with the sobering knowledge it would be almost impossible to recapture the same quality of ecstasy in another environment.

Nineteenth-century literature teemed with stories of the South Seas, many on the theme of maidens seduced by white men and then abandoned to regrets and longing. While she may have preferred that he stay, the girl who gave her love to a whaleman, who soon sailed away into the sunrise, was not really *abandoned.* She had her familiar island, her home, her family and her tribe. And her participation, as likely as not, was less great than the white lover fondly imagined. To him a crucially shattering experience, a surrender to illusion, he was not aware that to the dark-eyed, longhaired native, sex does not mean that much. It was imaginative literature which imparted the element of misery to the affair. A Polynesian *vahine* could not be made miserable because she had lived with an individual white *tane* (man). The Madame Butterfly theme was indicative of a literary trend which today happens to be in reverse. It is the American or European man who is in danger of being seduced by the Oriental woman whom he marries in frightening numbers.

The French historian André Maurois was disgusted with misinterpretations. "We owe" he said "to the Middle Ages the two worst inventions of humanity—romantic love and gunpowder." In chivalry, a man loves all but serves only one with the rule that he who is loved by a beautiful and virtuous woman carries with him a talisman that renders him invulnerable. In this conception, a spark of love saves a man from depravity. To love on impulse, to succumb to corporeal charms, goes the teaching of the West, is lust and weakness. In his wisdom Rochefoucauld admitted that "hearts still agitated by the memories of a passion" were more ready to receive a new one than when entirely cured. To an uninitiated young whaleman, who entered upon his maturity in the South Seas, it was the precious talisman he carried with him for his remaining days. For him, the instinctive quenching of the hot, fierce flame, in the first entrance into the mysteries, was both a gift and a sacrifice.

Human nature being what it is, there can never be one set of rules for the way of a man with a woman and differing codes will always be a matter of words. One thing is sure: The Tahitian is dedicated to the joy of living. He is the Etruscan of the South Pacific, drinking, feasting and loving. A *tamera* (banquet) lasts for hours on end with singing and ribaldry. With the invasion of the white man much of Paradise has been lost. *The Beach* typifies white commercialism. Much of the *Old Way* has given way to it.

Under the control of France, Tahiti's capital city of Papeete has its Grand Hotel with a roof-top restaurant where a *vahine* with waist-length black hair, golden skin and a pink hisbiscus flower behind each ear, serves the visitor paté, lobster, artichoke, Camembert, a bottle of Champagne in the best tradition of *la cuisine Française*. In Moorea or Tahiti a man today can still be a beachcomber, if he wishes, and live on less than a dollar a day which buys bread, avocados, bananas, pineapples.

While the *Old Way* has almost vanished under the encroachment of civilization, the flavor of the islands is still there. The South Seas remain the warm, glowing isles, caressed by soft winds, soothed with music, the scent of flowers, and blessed leisure. Peoples do differ in beliefs and behaviour in different parts of the world and the difference lies in what the Romans called *mores*. They cannot understand our attitudes and we cannot understand theirs.

Longfellow's poem *My Lost Youth,* which mourns for something never to be retrieved, is an example of the complicated thinking of the Westerner. He looks wistfully to the days when he was young, with the urge to recapture and repossess his youth. In the South Pacific, life is more simple. The Polynesian regards youth benignly and never expresses the wish to be young again. In a closer kinship with reality, he knows each phase of his life is part of a pattern. He moves with grace from one to the other. The older person looks with amused tolerance on the capers of the young with no envy, no curiosity, no impulse to inhibit or reform. Vicariousness is unknown in their philosophy. A native lives fully and richly according to his code and has no remorse, no regrets.

Herman Melville's sweetheart, the lithe and lovely Fayaway became the popular model of South Seas heroines—voluptuous, abandoned, her skin the color of *café au lait,* with a marvelous figure, her eyes glowing with deep tribal wisdom, her laugh low and throaty at times and at others a tinkling bell. It is not generally known, however, that Fayaway was a rarity among the native women, whose beauty lay not in their figures nor limbs, but in their skin and their lustrous black hair, their proud bearing and generosity. Renowned for his ruthlessly realistic portraitures of the Polynesians, Paul Gauguin painted the Tahitian woman as she really was—squat, short-jawed and big-footed.

By one of those oddities of evolution, there are women in the South Seas today who actually look more than ever like the beautiful Fayaway of Melville's heart. The beauty invented by the literary dreamers, more than a century ago, has become a living reality, the contribution of the seafarers who came in merchantmen, clippers, and lumbering whalers to leave behind their seeds and sometimes their hearts. When the Polynesians intermarried with Chinese, Australians, Portuguese, Americans and Japanese, it changed their physical structure. The female was transformed from a squat, pudgy and waistless build into a creature with elegant features, a molded jaw, longer body lines, slimmer legs and smaller feet.

A similar evolution affected the native dance known as the hula. Once it was a dreary, boring exercise endured for hours on end, performed to the beat of sticks and of drums made out of hollowed-out logs. Even the missionaries were impatient with its artlessness, but what really aroused their ire was the infrequent erotic courtship dance. This was more worthy of their condemnation, being a highly suggestive, blood-heating dance, often culminating in the public deflowering of a virgin by an adult warrior. During the years when the whalers were active in the Pacific theater, the hula had not yet evolved into the graceful dance we know, accompanied by the ukulele and Hawaiian ballads.

In retrospect, there were three arrivals important to the Garden of Eden. The serpent appeared in the guise of the white men Magellan and Cook, who first pierced the veil and brought in imports which began the corrosive destruction of Paradise. Second, on the heels of the explorers came the New England whalers, skippered by stern Puritans for the most part, or conscientious men who were intent on bringing a profit to investors at home who made the whaling voyage possible. Some of their crews were infected with ignorant and mutinous sailors. Third, the whalers were followed by men of another ilk—adventurers, profiteers, gamblers, and remittance men.

The first whalers to drop anchor in Honolulu arrived in Hawaii on September 17, 1819. There were two vessels, the *Equator,* Captain Elisha Folger out of Nantucket on her maiden voyage, and in her company the *Balaena,* Captain Edmund Gardner of New Bedford. From then onward, Honolulu was practically built on the whaling industry with the commerce of the whalers helping it to grow into a strong, modern city as Hawaii became a rendezvous for British, French and American trading ships.

In the middle of the nineteenth century, the Pacific Ocean

was the richest hunting ground for the Sperm and Right Whale. Shortly before the first half-century mark was reached, the American fleet of whalers comprised 680 vessels. Of this number, 640 were to be found cruising in the Pacific.

For a blissful thirty years the whalemen managed to keep the fruitful South Seas more or less to themselves. Then in 1850 came a holy invasion, which wreaked some havoc of its own. The missionaries, those unpaid, self-appointed police, arriving to bring God and clothing to the natives, disclosed to the world the whalemen's secret. The men wore black alpaca coats, high black shoes, and a sober-faced expression. Their wives hastened to put Mother Hubbards on the native women and to teach them some hymns. Their harsh morals, tinged with the rigidity of Cotton Mather's Massachusetts Bay Colony, their habitual lack of tact and grace were hardly in keeping with the Happy Isles.

From the very first, contact with the white men was devastating. Magellan and Cook brought in iron, gin and syphilis. Most people in the islands enjoyed perfect health and died of old age, their vigor sustained by the poi-fruit-fish diet which was nutritionally balanced. There was almost a total absence of bacteria and viruses. Captain Cook's men infected an entire chain of Kauai. In Fiji, the son of a royal chief returned from a visit to London with a small cough, and 30,000 Fijians died of measles. In Melanesia the common cold decimated whole populations. It was the same story over again as the succumbing of the American Indians to the white man's gifts.

The missionaries wrought havoc with the life and codes of the islanders. Some of the ideals introduced so ruthlessly among tropical people amounted to sheer madness. In the Marquesas Islands, for instance, they insisted that boys caught sleeping with girls had to marry them. This not only upset the ancient customs but increased the population beyond its ability to sustain itself. Besides it very much confused the Marquesans. The more sober of the missionary leaders were troubled over the impact of Christianity on Oceania and had to concede that "of the vices hypocrisy remains, of the virtues, gayety, simplicity and cordiality have disappeared." While the missionaries brought tuberculosis, altered sexual habits, a pervading sense of guilt, they also brought medicines, schools and the wheels of commerce.

The missionaries sent alarming reports back to the States that in Hawaii "intemperance and lust run riot with almost every anchored ship a floating brothel" as the Rev. Henry T. Cheever saw it. It was true that sooner or later most whalemen in the Pacific visited the grogshops and dancehalls of the Honolulu waterfront. They partook of the hospitality and entertainment of the natives on the scattered South Seas Islands. The comely native girls in their grass skirts were described by the missionaries as "lewd libertines" and their swaying beneath the palm trees to beating drums "a licentious dance which heated the blood." And the determined missionaries—called Misi—were here to stop it.

It was a pity. Whalers which arrived after the Misi took hold were no longer treated to the eyewarming sight of girls climbing up the anchor ropes in a tropical dawn to wake sleeping whalemen with their laughter and chatter. Instead, the heroes, who had risked death in contests with monstrous whales and the vicissitudes of the sea voyage, were welcomed decorously on shore by the faithful who in the name of Jesus Christ handed them a religious tract.

A few American missionaries came to Hawaii almost as early as the first whalers. While many were, to be sure, rigid fanatics, as far as the Hawaiian group is concerned, it must be said that they contributed the greater part to their development. By 1840, twenty years after they arrived, and a decade before the wave of missionaries hit other parts of the South Seas, they had established 375 schools on the Hawaiian Islands with an enrollment of 18,000 pupils. Aided by his associate, the Reverend

Asa Thurstan, Dr. Hiram Brigham made an important contribution. He reduced the Hawaiian language to writing.

And missionaries are still active in the whaling grounds of yesteryears. Today Elder M. Vernon Coombs zealously carries on the work of religion and education among Polynesians which he began some fifty years ago as a Mormon missionary from Utah. You will find the Mormons have built many schools and churches, and have a fine Mormon College of Liahona at Tonga.

CHAPTER

15

On the Emerald *to the South Pacific*

"Trouble is the next best thing to enjoyment; there is no fate in the world so horrible as to have no share in either its joys or sorrow.'

Longfellow (1807-82)

Stephen C. Phillips, one of Salem's most wealthy and respected merchants, sent out the Bark *Emerald* to the South Pacific. The log was kept by William H. Nichols, a cooper, and is perhaps the most outstanding of the Salem whaling journals. Fortunately for whaling annals, William Henry Nichols, son of the *Emerald* cooper, had the wisdom to submit his father's notebook to the Essex Institute. His letter dated March 24, 1913 to George Francis Dow, Secretary, was sent from his home at 347 Essex Street, Salem; I found it pasted into the front of the log. In a good firm hand, this is what he wrote:

With this letter I desire to transmit to you for the preservation in your archives, a journal of a whaling voyage to the South Atlantic, Indian and Pacific Oceans, on board the Barque *Emerald* of Salem, Captain Joseph Dexter, Master, and kept by my father, William Henry Nichols, Sr.

This voyage was begun on the 24th of May, 1838 and ended February 28, 1840. You will find in the front of this book a statement in my father's handwriting, which is dated March 25, either 1839 or 1889 probably. I found this among some of his papers and pasted it in this book. Also you will find in the front a copy of the statement which was in his handwriting which may be of interest, particularly where he describes the liberty which the Cook of the *Emerald* had on one occasion while on shore at Copang. Also in the front of the book you will find a clipping from the *Observer* giving an obituary which was published at the time of his death, dated April 7, 1902.

The author of this journal was born in the house still standing on Ash Street in Salem, October 20, 1816 and died in a house which he owned at the time of his death and built in 1851 at 12 Essex Street.

He married in 1843, Eliza Gauss, who was the daughter of Capt. Stephen Gauss, who was born in Trieste (Austria). Captain Gauss ran a sailing packet for several years between Salem and Boston. Eliza Gauss, my mother, was born in Salem February 19, 1824 and died November 20, 1860.

Scattered through this journal will be found probably a great many items which might interest anyone looking up the maritime history of Salem.

The last two or three pages of the Journal, he jots down some of his work in Salem after he returned from the whaling voyage and also a trip to Cienfuegos, saling from Boston, September 17, 1840 as a passenger on the Brig *Eliza Burgess.* Also another trip to the same place sailing from Boston on the Brig *Sarah Williams,* October 26, 1842. Also his return from Cuba, sailing July 6, 1843, on the Brig *Eliza Burgess.* On this voyage he was made a second mate at $20 a month.

On the back page you will see a drawing of the Barque supposed to be the *Emerald.* This was executed in lead pencil by myself October 24, 1858 when about 14 years of age. As I never saw the *Emerald,* of course this drawing is simply a fancy piece of work.

I think you will agree with me that this Journal is quite an important one and I trust it will be an acquisition to your present collection. The descendants of the author would like to have the pleasure (perhaps occasionally) of examining this Journal.

The parents of the author of this journal were William Nichols, Sarah (Millett) Nichols and are both buried in the Broad Street Cemetery in Salem, Massachusetts.

The journal presents a minimum of navigating detail, much good description of everyday life on board ship and in port, natural history and a whaleman's philosophy.

The title page of the journal lists the names of the *Emerald*'s officers and crew. A third of the number came from Salem, with the captain from Martha's Vineyard:

Joseph Dexter, Master of Vineyard
George W. Downes, 1st Officer Vineyard
William Stafford, 2nd Officer, Natick, R. I.
Isaac C. Daggett, Boatsteerer, Vineyard
John Collier, Boatsteerer, Salem, Mass.
David B. Edwards, Boatsteerer, Salem, Mass.
William H. Nichols, Cooper, Salem, Mass.
John H. Francis, Shipkeeper, Salem, Mass.
Henry A. Tuttle, Salem, Mass.
Ledyard Farnsworth, Salem, Mass.
Joseph Jackson, Salem, Mass.
Elbrige C. Upham, Carpenter, Salem, Mass.
Christopher Sherman, Natick, R. I.
Charles Ingalls, Lynn, Mass.
William Durant, Lynn, Mass.
Mathew Potter, Bangor, Maine
George W. Nichols, Bangor, Maine
Joseph Ravens, Bangor, Maine
Charles Hayes, Bangor, Maine
James Pico, Portugal
Manuel Flores, Portugal
Thomas Osborn Canacker
George W. Powell, Cook, Baltimore
Samuel Maury, Steward, Phil. Pa.

The only two logs I found which listed the names of the crew were the *Elizabeth*'s and the *Emerald*'s. As seen in a previous chapter, the ages were given by the log keeper of the *Elizabeth* and also the assignment of the men to the various whaleboats. It is possible with this information to prove that Salem men were skilled in harpooning and lancing whales, a highly specialized division of the whaling industry. The *Elizabeth*'s steerer was Abraham Anthony an experienced harpooner of New Bedford, 36 years old. But James Kimball, a 24-year-old whaleman of Salem is listed as *bow oarsman,* which also classifies him as a harpooner. A whaler's crew was divided into as many divisions as there were whaleboats, but there were always only two try-out watches each constituting half of all men on board.

As for the *Emerald*'s crew, we know that the captain was from Martha's Vineyard as well as the 1st Officer, George W. Downes. Apparently there were three harpooners. Isaac C. Daggett came from Martha's Vineyard, but the other two harpooners, John Collier and David B. Edwards, were Salem men. Other skilled seamen, such as coopers and carpenters, abounded in Salem, and they are numbered among the *Emerald*'s crew. One hardly thinks of Salem producing its own harpooners and lancers, but the records furnish indisputable testimony that Salem could and did fulfill all the requirements of a whaling cruise many times over.

The *Emerald* departed from the Stephen C. Phillips Wharf in Salem on May 24, 1838 for the whaling bays of New Holland, and her voyage was planned for the South Atlantic, Indian and Pacific Oceans.

Two months and more from her departure in Salem, the *Emerald* approached the Cape of Good Hope, at the tip of South Africa, on August 10th, and got around from the South Atlantic into the Indian Ocean whaling grounds. There was good weather and high living from the last four pigs slaughtered for the pantry. Finally on August 26th an entry appears in the log that whales have been sighted. The boats were lowered, but no whales were captured. By September 24th—four months out of Salem—still no whales. A great deal of the provisions was gone without a penny of profit. The officers and crew were desolate. The Indian Ocean was shark-infested which made for perilous fishing. The long spell of bad luck was broken when the first whale, a Sperm, was taken. This was followed closely by the capture of another on October 3rd, the oil from the first two whales had not yet been put into casks as more whales were expected. Two weeks passed and on October 16th as the *Emerald* cruised off Cape Lewen, New Holland, she took a forty-barrel whale alongside.

This success elevated the men's spirits. Three days later they were further heartened when 80 barrels of oil were secured. A two-week rest followed before whales were again sighted. Getting a slow and discouraging start, the *Emerald* enjoyed good fortune and within three months of activity secured a cargo of 1,100 barrels of whale-oil and 218 barrels of sperm. From October 19th to February 22nd there were no more whales, then came a chase long to be remembered. The crew put up a gallant battle to fasten to "an ugly customer" who refused to be captured. This whale counter-attacked. The whaleboat was overturned and the crew hung on to the bottom for their lives as they watched the prize swim out of sight. The starboard boat picked up the drenched men and took them back to the *Emerald.* In addition to losing the whale, much equipment was lost, too. The whale carried away with him five irons, one lance, one spade, short warps. Three boat paddles were gone, one whole line and part of the other tow lines.

In addition to securing and boiling whales, the balance of the *Emerald*'s log reveals the doings of George W. Powell, the Negro cook from Baltimore, who was so proud of his ancestry, and kept the ship amused with his original songs and stories; also the doings of Samuel Maury, the Negro steward from Philadelphia. There is a description of an invasion of mice which damaged the sails; a visitation of dysentery and scurvy, and finally getting provisions during a Malayan war in which the *Emerald*'s crew saw many native villages burned. There is an entry which tells of receiving news that Captain Wilkens of Salem was cut off on the Coast of Sumatra by the Malays in his Ship *Eclipse.*

There follows the happy entry of high living on fresh food

after the interminable months at sea; securing oil after a wait of three months; descriptions of fighting between the French and the English, more quarrels on board between the officer of the deck and Captain Dexter; scarcity of fire wood, but abundance of provisions, the visitation of tormenting fleas, and finally the scarcity of the ship's provisions "with nothing left but bread and still 944 miles from Salem."

When the *Emerald* met up with the Ship *Naples,* Captain Henry Archer of Salem, there was great rejoicing. The *Naples* was 194 days from China and 40 days from St. Helena, bound for Boston. The boats were lowered for a gam and Captain Dexter in the best of whaling tradition stood upright while his men rowed him over to the *Naples.* William Nichols, the logkeeper, was among the "number of Salem boys" who eagerly clambered up to the deck of the *Naples,* where he found two of his old acquaintances. There was little news of home as the *Naples* had been absent from Salem for 16 months. The "acquaintances" were William Edwards, 2nd Officer and Henry Archer, the son of the Captain.

By February 21st, the *Emerald* had reached Long Island, New York, by the grace of a strong stiff breeze which propelled the ship forward through a thick fog in which she found the Pilot Boat *Superior* of Martha's Vineyard at 4 in the afternoon. From the *Superior,* the *Emerald* secured a pilot, Mr. Dexter, who was an uncle of the *Emerald*'s Captain. By his reckoning, the colored cliffs of Gay Head, Martha's Vineyard, were only 38 miles distant. The *Emerald* had made Cuttyhunk light and soon the Gay Head light, and by 9 A.M. sailed through the Sound and stood into the shores of Martha's Vineyard "as near as convenient to Captain Dexter's residence." Three guns were fired within hearing distance of the Captain's lady, the first notice that her husband's voyage was nearing its end.

By 11 o'clock the *Emerald* was at anchor in Holmes Hole Harbor in company with a number of other vessels. Friends of the Captain, the 1st Officer and Isaac Daggett, boatsteerer, who were all residents of the Vineyard, came and took the lucky whalemen ashore. The crew which remained on the *Emerald* were described by Nichols as "us poor fellows on board who were left to worry off a head wind."

It is easy to guess the sentiments of the returning whalemen. The voyage had been successful. They had seen hard times. Captain Dexter and Daggett went on shore to be welcomed by their families and to enjoy once more the warmth of their households. The balance of the crew were both hungry and terribly excited about nearing home. Since January 26th, or for a month, there had been an end to the ship's molasses and "almost everything eatable with the exception of bread." Captain Dexter did not overlook this. As soon as possible, he sent on board the *Emerald* "some wood, beef, pork and sugar." The supplies arrived at sundown. The Captain and Mate, it appears, remained at their homes on Martha's Vineyard from Saturday, February 22nd to Monday, February 24th. As the weather was more clear at sun-up on Monday than it had been the previous two days, it was possible to continue the last lap of the voyage to Salem Harbor. The *Emerald* fired a gun as signal to the Captain and Mate. They came along and brought a Pilot on board to guide them through the rocks and reefs of the coast line.

At sunrise the *Emerald* was under way. It was a pretty sight for she departed in company with seventeen other vessels, leaving for various ports. By noon, the *Emerald* was tacking off Nantucket Island and came to anchor for the night in Edgartown Harbor. At 4:30 in the morning of Friday, February 28th, a light was seen which was taken to be a Pilot boat's light. The *Emerald* was steered toward it with the signal lanthorn set and the mate also burnt a blue light. They soon found it was the revolving light of the Boston lighthouse. Soon after 5 A.M. they caught sight of the beams of the lighthouse on Baker Island and Marblehead. It was very cold. At 7 A.M. the Pilot boat *Vigilance* spoke the *Emerald* and her Master, Pilot Perry, was taken on board. Baker Island was now only two miles distant. At 8:30 the *Emerald* came to anchor in Salem Harbor, February 28, 1840.

Soon a large number of boats rowed out and surrounded the *Emerald.* In them were the friends and relatives who could not wait for the *Emerald* to dock at the wharf. "On landing," wrote Nichols, "we found a great many of our friends on the wharf to receive us, but as we passed off of the wharf, we nearly all separated, some going one way and some another and thus ends this voyage after being gone 21 months and 3 days, returning the whole crew 24 in number and hailing 1,700 barrels of right whale-oil and 300 barrels of sperm oil."

The log contains a detailed list of whales struck and saved, and those lost as well as the names of the men in the boats getting the whales. For instance, during October, 1838, five trips were made in boats to get whales by three different boats, one of which was headed by Captain Joseph Dexter. During October, 4 Right Whales were lost and 5 Right Whales and 1 Sperm Whale were saved. Also lost were 2 Sperm Whales. One boat was sunk. Evidently, the logkeeper felt that this was a good batting average for the entry concludes with the words: "So here is success to the whalemen evermore!" written in beautiful penmanship with a gay flourish.

Some excerpts from Mr. Nichols' log:

August 10, 1838. Toward the Cape of Good Hope.

Sunday: Good weather for the latitudes, shooting sea birds . . .

Monday, Aug. 20. Good weather. The remaining four pigs we had have been slaughtered this day. We live high . . .

Aug. 22. Caught one albatross being very small only measuring eleven feet from the tips of its wings and saw a great many around the ship.

Aug. 26. First moderate weather. Had pleasure of seeing Porpoises under our bows for a few minutes at half past 6 A.M.

Mr. Dowenes being at masthead saw Right Whales, two of them. Got breakfast; hauled the main yard aback and lowered our three boats and pulled for them. After pulling for them and going on to them several times, we could not get fast to them. (The *Emerald* left the Phillips Wharf in Salem in May. These are the first whales seen. Their capture was unsuccessful.)

September 24, Still no whales. (There are no other words on this page but the bleak ones noted) Barque *Emerald,* Dexter of Salem, four months out. No oil.

September 25. (The *Emerald* is now in the Indian Ocean, gets its first *fish* a Sperm Whale. Three boats are put in pursuit and the school went off with several spouting blood) Took two whales alongside at 9:A.M. and went to boiling 5 P.M.

Wednesday. Oct. 3d. . . . Last night Mr George W. Powell's office doors were broken open and entered which hurt the poor darky's feelings. (Powell was the *Emerald*'s Negro cook from Baltimore, Maryland.)

Oct. 4th. Thick, dirty weather. . . . We have the oil of our two Sperm Whales on deck yet waiting for more before stowing down.

Oct. 16. Got a forty barrel whale alongside. (By this time the *Emerald* was cruising off Cape Lewen, New Holland . . . spoke whale ship *Huntress,* Capt. Hull of New Bedford and found they were out six months and had 300 barrels of Sperm and 200 bbls of whale-oil)

Oct. 19. 80 barrels of oil. (There now follows a day by day record of Ledyard Farnsworth of Salem who fell from the masthead and was almost killed. His condition and appetite are reported in spite of tremendous whaling activity of chasing and boiling down. It appears Farnsworth fell in reefing the topsail on Oct. 22.)

Nov. 1. Chasing whales. Henry A. Tuttle jammed his little finger bad.

Nov. 8. Ledyard Farnsworth gets along slowly. . . . Fine times. Whales plenty.

Nov. 14. Farnsworth is on duty and Tuttle is getting along.

Dec. 24. Barque *Emerald* seven months out. 1100 barrels of whale-oil and 218 barrels of sperm-oil.

Jan. 15, 1839. Joseph Jackson has been crazy this day.

Thursday, Feb. 21. Moderate weather. Saw nothing. For Sperm Whale we are cruising.

Feb. 22. (This entry tells of pursuit and overturning of whale-boat with crew hanging on to the bottom, and the whale out of sight. The starboard boat picked up the drenched crew and took them to the ship.) Lost this day by the whale five irons, one lance, one spade, short warps, three boat paddles and one whole line and part of the other tow lines. (In large letters with flourishes—) DISCOURAGING.

Mar. 2. Fine weather. Sighted and chased whales. Ends with no sperm this day. Cook and steward had a battle amongst themselves.

Mar. 17. Overrun with mice which did damage to sails in the sail room.

Mar. 23. Out ten months 1100 bbls of right and 260 bbls of sperm-oil.

Mar. 27. Got whales and started boiling at midnight. Plenty of sharks around this night. We are off NorthWest Cape of New Holland.

April 6. Fine weather cruising in company with the *Wade.* At 3 P M saw another sail ahead; at 5 P M lowered our Starboard boat and gammed with the *Wade.* At 7 P M spoke the stranger and found it the *Clifford Wayne,* Capt. Downes. Went on board *Wayne.* At midnight each boat's crew went on board of their respective ships. From the Barque *Wade* we got 12½ lbs. of rivets, saw nothing. PLENTY OF COMPANY

April 11. Most of the crew complains of the Dysentery.

April 29. Joseph Ravens is swelled considerable with the scurvey, face and feet.

May 4. Captain Dexter went on shore while ship was lying at anchor in Ampanan Bay and sent off three dozen ducks, two very fat hogs and a few sweet potatoes and squashes . . . Malays still fighting. Two or three villages burnt this day. *Emerald* here to recruit.

May 5. Fine weather. Thanksgiving day with us fore and aft. Living very high on fowles and vegetables, beef, etc.

May 6, 1839. War at an end. King of Bally town is killed and Ampanan King now will rule. (The *Emerald* whalemen celebrated the Ampanan victory by setting off a number of sky rockets, blue lights and Roman candles, while the ruling party hoisted the Dutch flag.)

May 10. Heard by Ship *Morrison* of New York that the Ship *Eclipse,* Captain Wilkens of Salem had been cut off on the Coast of Sumatra by the Malays and most of the Ship's company massacred. . . . Ship still recruiting.

May 11. Good living on fowls, fruit. Little too much for Charles Ingalls who came aft with his trousers off and gave them to the Captain, telling him that they were too big for his hands. Ingalls sent forward for the night.

May 12. Charles Ingalls put in irons again this morning, considerable conversation passed between Captain and Ingalls. The former read a part of the ship's articles to the latter, but not much notice did he take of it. However, Ingalls was ordered to keep aft and sleep in the steerage.

May 15. Ingalls is yet in irons and Captain Dexter on shore for the night.

May 17. Dexter on shore for the night.

May 20. Joseph Ravens gets better fast.

May 21. (This includes a long account of provisions taken on board and about the Ship *Morrison* to sail in two days for Singapore and Canton with a cargo of rice). Myself and a part of our ship's company put letters on board the *Morrison* to be forwarded to the States first opportunity. Joseph Ravens and Portugal Jim yet unwell. Charles Ingalls is yet in irons.

AT SEA ONCE MORE

June 24. Thirteen months out and have not taken a whale to the ship for nearly three months. We are at South Coast of New Holland in Doubtful Island Bay.

July 4. Stowing down oil on the 4th of July. Who would have thought of such a thing!

Saturday, July 13. RIGHT WHALES IN DEMAND ON BOARD THE BARQUE EMERALD.

July 23. Whales scarce.

Aug. 4. Looking sharp for whales.

Aug. 9. French and English are fighting.

Sept. 25. Good whaling.

Sept. 26. Barque *Emerald* wanting but a few more barrels of oil.

Oct. 5. Homeward bound! All full!

Dec. 10. Trouble on board over mending a topsail. Some very hard words were spoken between Mr. Downes, officer of the deck, and Captain Dexter . . . All of this trouble arose from that old fore topsail concerning the manner of mending it. . . . At breakfast table there was another row between the Captain and Mr. Downes. CAPTAIN AND MATE QUARRELING. *Emerald* off Cape Horn full of oil.

Dec. 20. This day we commence on having only one warm meal a day and that being dinner. This is owing to our being short of wood, but having plenty of provisions, bread, water, etc.

Dec. 25. The decks were scraped and preparations made to receive any whales if caught.

Jan. 6, 1840. Ship seems to be infested by a troublesome insect commonly called fleas.

Jan. 21. We are living on temperance principles. For breakfast and supper cold water is used as a substitute for hot coffee and tea. Our wood is almost gone.

Jan. 26. We have found an end to the ship's molasses and almost everything eatable with the exception of bread. (By this time the log pages have flourishing letters on top—TOWARDS SALEM)

Feb. 10. Salem is 944 miles distant.

On February 28, 1840, the *Emerald* anchored in Salem. She reported with 1700 barrels of Right Whale oil and 300 barrels of sperm oil. The share of Mr. Nichols was 1/50th of the voyage which amounted to $559.38 for a year and nine months' service. Mr. Nichols recorded that whale-oil sold for 30¢ a gallon, sperm oil at $1.00 a gallon. Each barrel held about 30 gallons, so there were 51,000 gallons of whale-oil worth $15,300, and the sperm was worth $9,000 for only 9,000 gallons. As Mr. Nichols mentions whalebone being worth 10¢ a pound, it is possible that the whalebone was disposed of before the *Emerald* returned to Salem.

William H. Nichols, the log keeper and senior cooper, went whaling a few days before he was twenty-one. Like many boys of his era, his responsibilities were assumed at an early age. His father died when he was twelve, leaving William the eldest boy in a large family. To give assistance, he left school and hired out to a farm in Ipswich, then to a farm in Beverly. It was not unusual in large families for youngsters to be doing a boy's work by five and a man's work by the time they were ten. Work often passed for play. When a boy went swimming, he took his rake along to get clams. The places chosen for a picnic were cranberry bogs or a blackberry patch. The morning sun never shone on any Yankee boy in bed. At fourteen, William was apprenticed for seven years to a man named Savory to learn the cooperage trade. This training stood him in good stead. After his whaling voyage, he went to work as a cooper in Cuba, then set himself up in this business upon his return home to Salem.

The whaling voyage on the *Emerald* awakened in Mr. Nichols an interest in the whaling industry which he never lost and in after years he invested in several whalers. From notes inserted into the log book, we find that Mr. Nichols served as bow oarsman to Captain Dexter. According to whaling form, this signifies that he had his chance at darting the harpoon. From these informal notes, we learn that the Captain of the *Emerald* was so gratified by completing the voyage before expectations and returning with a full ship and all hands in good health, that he had mapped out another voyage. The owners, however, did not see fit to send him and Captain Dexter never again sailed on a Salem whaler. "The *Emerald*," declared Nichols, "sailed under another master and made a very poor voyage."

Nichols' statement was no doubt based on a deep loyalty and even affection which he felt for his Skipper. It was disappointing that after such a successful voyage, young Captain Dexter was not again commissioned to command the *Emerald*. Sailing for Stephen C. Phillips in 1840, the *Emerald* was under Captain Robert Brown, a Salem man. From the customs records we can piece together the real story. In 1840 under the new master the *Emerald* actually made a good voyage, but not as good as she had achieved under Captain Dexter. The *Emerald* returned from the Indian Ocean whaling grounds with 400 barrels of sperm oil, 1,100 barrels of whale-oil and 8,800 pounds of whalebone. The value of her cargo was roughly $24,000. The time consumed was 33 months.

To compare voyages. The *Emerald* under Captain Dexter returned from the same whaling grounds with 250 barrels of sperm, having disposed of 100 barrels ahead of time, and 1,750 barrels of whale-oil. The oil was worth about $26,250. No whalebone is mentioned in the customs report. However, whale-oil from the Right Whale means baleen or whalebone, too, and the 1,750 barrels of whale-oil meant at least $2,000 worth of bone, or a total cargo of over $28,000. The time consumed was 21 months, or a year less than Captain Brown's subsequent voyage in the same whaler. Therefore, the *Emerald* made a swifter and more productive voyage under Captain Dexter than she did under Captain Robert Brown. On the strength of the record, however, Captain Brown's voyage cannot be considered "a very poor voyage" as Nichols mentions in his notes.

Youth and whaling were synonymous. It was a calling which required enthusiasm and physical vigor, often tempered by the wisdom and eternal youthfulness of whaling captains, whose energies belied their years. Captain Dexter was only 28 years old when he took out the *Emerald* in 1838. The vessel's company was under that age except two or three who were considerably older. The boatsteerers or harpooners were about 23 years old and had already attained considerable whaling experience. "They seldom missed fastening to a whale with a decent chance" Nichols recalled. "The foremast hands were quite young too, mostly green hands, who soon got broken into the business." The older men on the voyage were the cook and the steward, the aged darkies, who added music and cheer to the 1838 voyage of the *Emerald*.

CHAPTER

16

On the Reaper *and* Izette

"In this world of change naught which comes stays and naught which goes is lost."

Madam Swetchine (1782-1857)

I found twenty-three logs (in the collection at the Essex Institute) in two wooden boxes in the basement. They were dusty and worn, some 120 years old. No doubt more logs will come in with the dissolution of private libraries or estates. Deciphering and handling a sea journal of a whaling voyage is almost as good as being there. Copying from original logs is something of a feat in itself. The ink has faded. The writing is often poor and cramped. Many pages are blurred with oil spots and unfortunately logs are incomplete, parts separated and lost.

Invited by the Director to take my pick of the helter skelter, unclassified relics, I chose the cleanest, best preserved and most legible only to find they were very dull. The notebooks whose entries were barely decipherable, and in a miserable condition, were the real finds. With a sense of intruding upon the secrets of Neptune's locker and with a deliberate intent of salvaging these lost literary gems from further oblivion, I painfully copied several of the journals. The contents were published when this book appeared in its first edition in 1951.

Captain Neal kept a good journal of the Bark *Reaper* on its whaling voyage to the Atlantic and Indian Oceans, which speaks for itself.

Aug. 21, 1837. At 3:00 P.M. left Derby's Wharf Salem with the wind from N.E. and a thick fog bank to the East, stood under the north shore and came to in a thick fog.

Aug. 22. At 12 M got under way at 1 P M past Baker's Island from which I took my departure and steered E by S with the Wind W.

Aug. 23. All these 24 hours brisk breezes from W, steering E ½ S Latitude 42 27 N, Long. 67 40 W

Aug. 24. All these 24 hours moderate breezes from S.S.W steering E.S.E past English ship, steering E N E.

Aug. 25. First part of these 24 hrs calm, latter part fresh breezes from E N E, steering S.E. a ship to windward steering prevailed with us.

From the Cape de Verde toward the Line—18 Sept. 1837.

Friday, 22. 5:00 A.M. *Fogo* in sight 39 miles distant, baring N W, the *Heroine* in sight, calm at 10 light airs from N, steering S by W at 8 A.M.; took a black fish in good time as our oil was rather short.

Sat. 23. All these 24 hours gentle breezes from N, steering S by W. Boiled the black fish made about 2 bbls black as the . . . . (undecipherable).

Wednesday, 27. All these 24 hrs faint airs from W S W steering on the wind to the S and E. picked over the potatoes, found some were rotten.

Tuesday, Oct. 2, 1837. First part of these 24 hrs faint airs from S W, latter part strong breezes from S W to S S W with strong squalls of rain at 2 P M. Spoke the *Friendship* west of Fairhaven, 50 days out. Capt. W. told me that he has crossed the equator and Line at the same time and made a short passage in doing so I find there are many of the whale ships cross in from 18 to *0*, then take the S E trades. They fetch well to the eastward.

Monday, 9. Crossed the Line. Fresh trades and pleasant steering

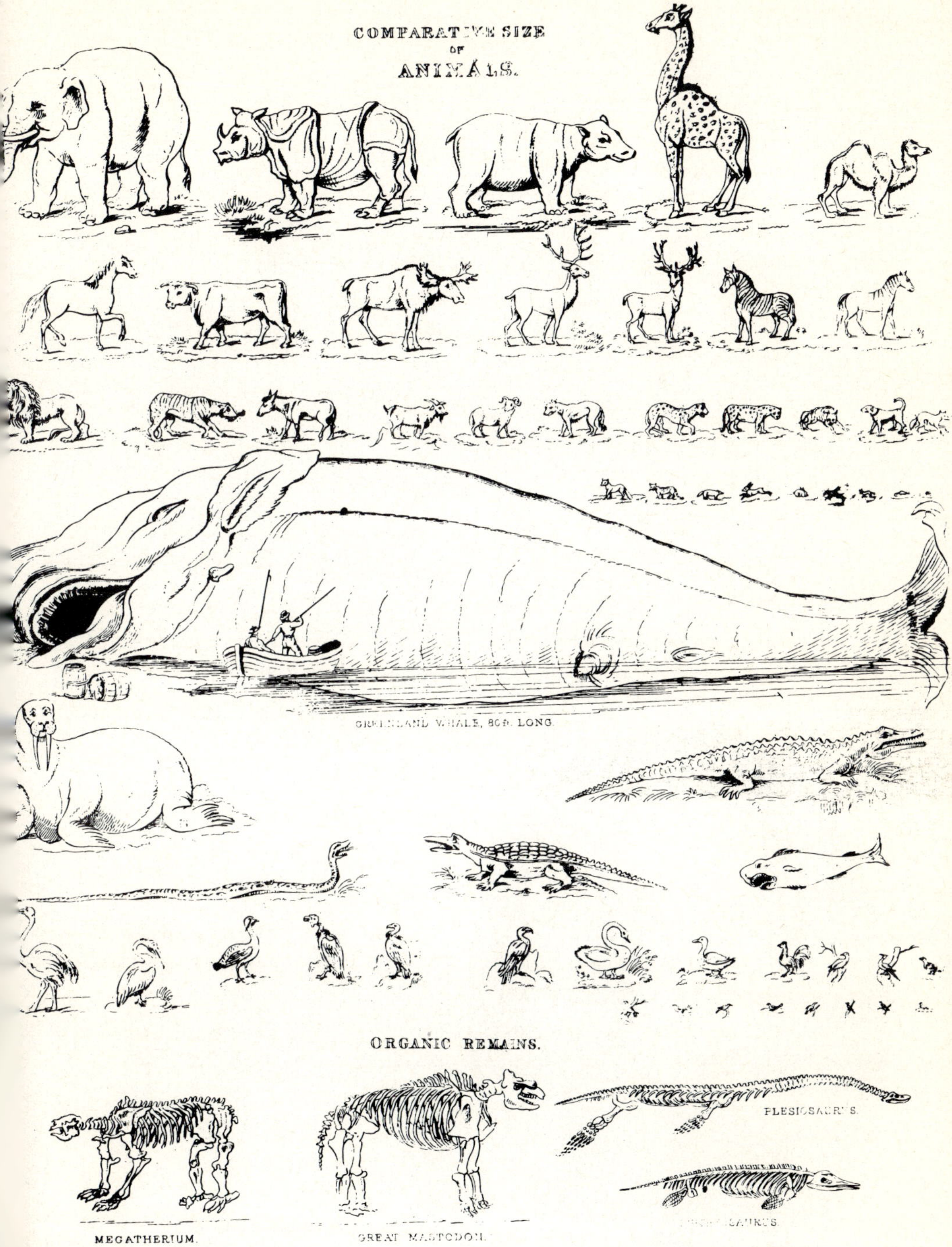

The whale compared to other animals from an old drawing by W. Croome engraved by J. Andrews. Very rare old print.

Arctic whaling scene after Currier & Ives from oil painting by Natalean and Noble Cassady of California.

Metal and glass whale-oil lamps. Essex Institute, Salem.

Captain William Ramsdell (1766-1842) of Salem, Master of the Salem Ship Lydia. *Taken from the album of the Salem Marine Society. Kindness of the late Colonel Arthur Treadwell Dalton, a Salem member.*

Arctic perils. The fleet enclosed in the ice-pack.

The camp on a grounded iceberg.

Escape of the Bark Florence. *"Recent disasters of the whaling fleet" by Captain W. H. Kelley. Harper's Weekly, November 18, 1876.*

Eskimos whaling off-shore and fishing by spearing under the frozen surface. By Roberto Freschi in an ancient Italian volume. Author's collection.

The Bark Wanderer *which was wrecked on the Island of Cuttyhunk on August 25, 1924, having sailed from New Bedford two days earlier. Shown on the whaling grounds—without the usual white waist band—by the famous marine artist Gordon Grant. Collection of The Seamen's Bank for Savings, New York City.*

Sectional view of English whale factory ship shows, on boat deck: (1) aircraft catapult, (2) hangar roller shutter, (3) port hangar, (4) aircraft, (5) fans, (6) aircraft workshop, (7) fans, (8) port funnel uptake, (9) fans. On the halfdeck: (10) aircraft fuel tanks, (11) hospital, (12) bathrooms, (13) doctor's consulting room, (14) doctor's cabin, (15) dental surgery, (16) dental surgeon's cabi.., (17) gangway, (18) assistant surgeons. On the flensing deck: (19) cookers' cabins, (20) derrick, (21) hatches to pressure boilers, (22) blacksmith shop, (23) carpenter shop, (24) derrick and funnel, (25) lifeboat, (26) 40-t winch, (27) electrical control room, (28) galley, (butcher shop, (30) pantry, (31) bakery, (32) hat (33) stewards' store, (34) boatswain's store. upper flat: (35) seamen's laundry, (36) stewar laundry, (37) funnel uptake, (38) bone drying pla (39) blubber boilers, (40, 41 and 42) process tanks. On the factory deck: (43) stern entrance slipway, (44) whale being drawn up slipway, (cabins for workers, (46) drier, (47) meat meal pla (48) separators, (49) liver plant, (50) mixture tan

The British Whaling factory-ship Balaena *active in the Antarctic. Drau*

51) pressure boilers, (52) conveyor, (53) driers, 54) settling tanks, (55) separators, (56) cookers, 57) cutting machinery, (58) driers, (59) freezing ompartment, (60) hatch, (61) harpoon stores, (62) hain lockers, (63) stores. Lower flat: (64) fresh ater tanks, (65) knife sharpening room, (66) tank ops. Lower hold: (67) steering gear, (68) water allast tanks, (69) rudder, (70) propeller shaft, (71) riple expansion steam propulsion engine, (72) esh-water tanks, (73) boiler room, (74) generators, 75) fuel tanks, (76) Diesel oil tanks, (77) settling tank, (78) fuel tank, (79) pump rooms, (80) forehold, (81) fore deep tanks, (82 and 83) fresh-water or ballast tanks, (84) reinforced bow frames. Bridge and superstructure: (85) foremast, (86) beam radar, (87) master compass, (88) wireless cabin, (89) captain's chartroom, (90) wheelhouse, (91) officers' cabins, (92) dining room, (93) officers' and chief stewards' cabins, (94) mess and smoke rooms, (95) hatch, (96) windlass. This cutaway drawing was made of the new Balaena, 15,000-ton whaling ship built in Belfast for United Whalers Ltd. of London.

Davis, long famous for his accurate marine art studies, Chief Staff Artist of the Illustrated London News.

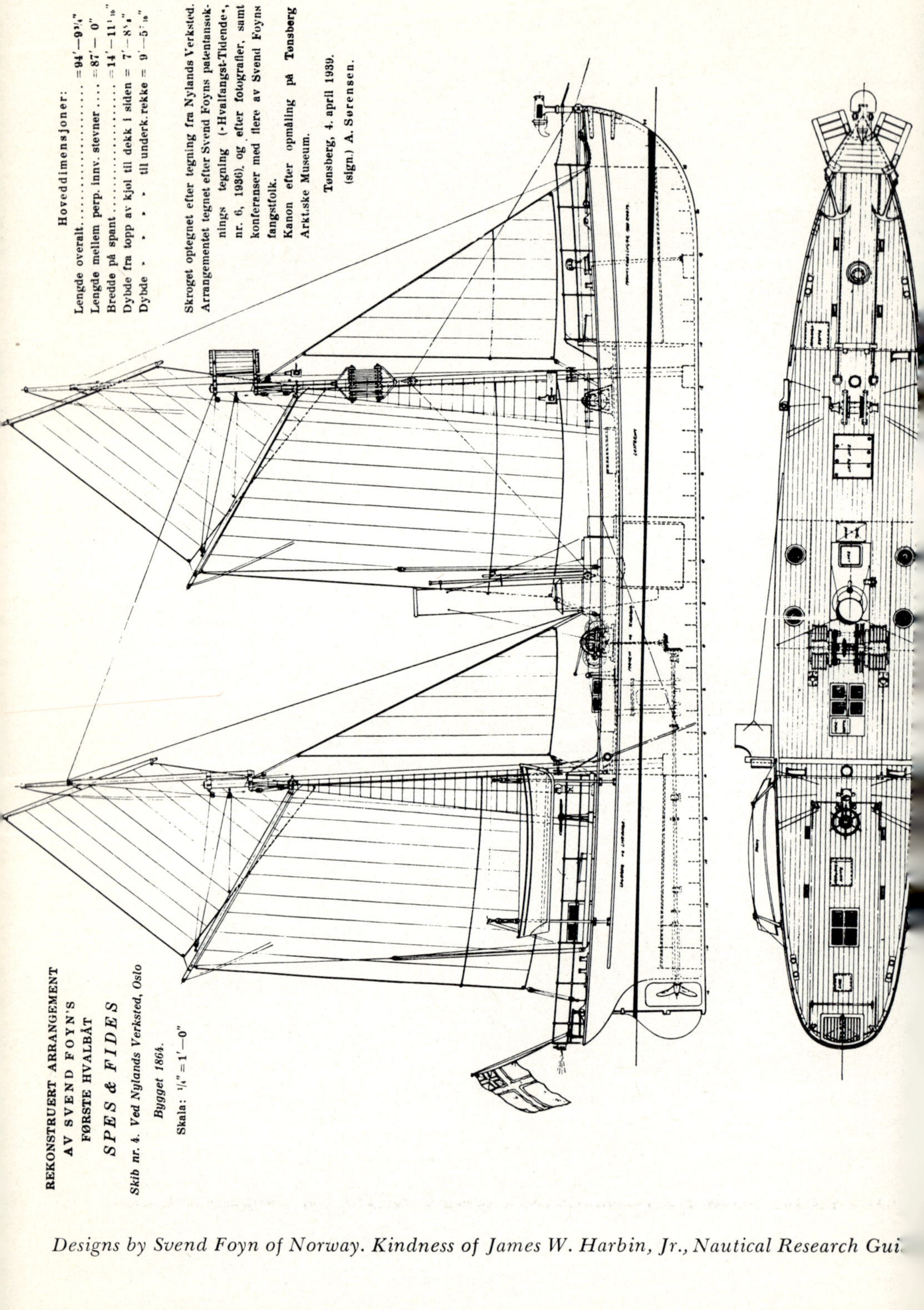

Designs by Svend Foyn of Norway. Kindness of James W. Harbin, Jr., Nautical Research Gui

Salem Custom House in 1850 at the head of Derby Wharf.

Benjamin Webb—Salem whaling agent.

Nathaniel Hawthorne's stencil for marking merchandise at the Salem Custom House where he conceived his great novel, The Scarlet Letter.

Holder for spermaceti candles.

Whale-oil lantern.

Front doorway of the Pingree House on Essex Street and elegant hall way.

Whaling Ship try-works, Nantucket.

The old Candle Factory on Nantucket Island.

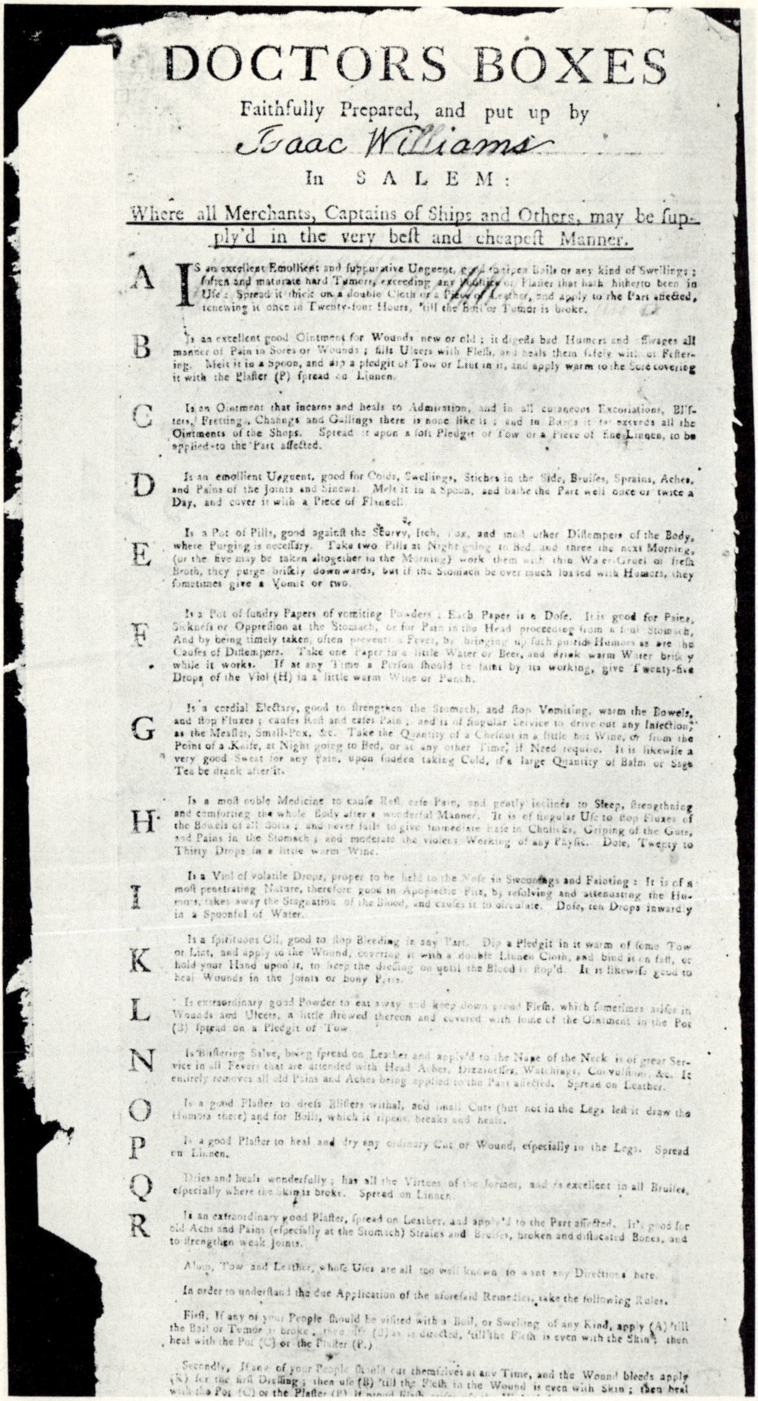

DOCTORS BOXES

Faithfully Prepared, and put up by

Isaac Williams

In SALEM:

Where all Merchants, Captains of Ships and Others, may be supply'd in the very best and cheapest Manner.

A — IS an excellent Emollient and suppurative Unguent, good to ripen Boils or any kind of Swellings; soften and maturate hard Tumors, exceeding any Poultice or Plaster that hath hitherto been in Use: Spread it thick on a double Cloth or a Piece of Leather, and apply to the Part affected, renewing it once in Twenty-four Hours, 'till the Boil or Tumor is broke.

B — Is an excellent good Ointment for Wounds new or old; it digests bad Humors and asswages all manner of Pain in Sores or Wounds; fills Ulcers with Flesh, and heals them safely without Festering. Melt it in a Spoon, and dip a pledgit of Tow or Lint in it, and apply warm to the Sore covering it with the Plaster (P) spread on Linnen.

C — Is an Ointment that incarns and heals to Admiration, and in all cutaneous Excoriations, Blisters, Frettings, Chafings and Gallings there is none like it; and in Burns it far exceeds all the Ointments of the Shops. Spread it upon a soft Pledgit of Tow or a Piece of fine Linnen, to be applied to the Part affected.

D — Is an emollient Unguent, good for Colds, Swellings, Stiches in the Side, Bruises, Sprains, Aches, and Pains of the Joints and Sinews. Melt it in a Spoon, and bathe the Part well once or twice a Day, and cover it with a Piece of Flannel.

E — Is a Pot of Pills, good against the Scurvy, Itch, Pox, and most other Distempers of the Body, where Purging is necessary. Take two Pills at Night going to Bed, and three the next Morning, (or the five may be taken altogether in the Morning) work them with thin Water-Gruel or fresh Broth, they purge briskly downwards, but if the Stomach be over much loaded with Humors, they sometimes give a Vomit or two.

F — Is a Pot of sundry Papers of vomiting Powders: Each Paper is a Dose. It is good for Pains, Sickness or Oppression at the Stomach, or for Pain in the Head proceeding from a foul Stomach, And by being timely taken, often prevents a Fever, by bringing up such putrid Humors as are the Causes of Distempers. Take one Paper in a little Water or Beer, and drink warm Water briskly while it works. If at any Time a Person should be faint by its working, give Twenty-five Drops of the Viol (H) in a little warm Wine or Punch.

G — Is a cordial Electary, good to strengthen the Stomach, and stop Vomiting, warm the Bowels, and stop Fluxes; causes Rest and eases Pain; and is of singular Service to drive out any Infection, as the Measles, Small-Pox, &c. Take the Quantity of a Chesnut in a little hot Wine, or from the Point of a Knife, at Night going to Bed, or at any other Time, if Need require. It is likewise a very good Sweat for any Pain, upon sudden taking Cold, if a large Quantity of Balm or Sage Tea be drank after it.

H — Is a most noble Medicine to cause Rest, ease Pain, and gently inclines to Sleep, strengthning and comforting the whole Body after a wonderful Manner. It is of singular Use to stop Fluxes of the Bowels of all Sorts; and never fails to give immediate Ease in Cholicks, Griping of the Guts, and Pains in the Stomach; and moderate the violent Working of any Physic. Dose, Twenty to Thirty Drops in a little warm Wine.

I — Is a Viol of volatile Drops, proper to be held to the Nose in Swoonings and Fainting: It is of a most penetrating Nature, therefore good in Apoplectic Fits, by resolving and attenuating the Humors, takes away the Stagnation of the Blood, and causes it to circulate. Dose, ten Drops inwardly in a Spoonful of Water.

K — Is a spirituous Oil, good to stop Bleeding in any Part. Dip a Pledgit in it warm of some Tow or Lint, and apply to the Wound, covering it with a double Linnen Cloth, and bind it on fast, or hold your Hand upon it, to keep the dressing on until the Blood is stop'd. It is likewise good to heal Wounds in the Joints or bony Parts.

L — Is extraordinary good Powder to eat away and keep down proud Flesh, which sometimes arises in Wounds and Ulcers, a little strewed thereon and covered with some of the Ointment in the Pot (B) spread on a Pledgit of Tow.

N — Is Blistering Salve, being spread on Leather and apply'd to the Nape of the Neck is of great Service in all Fevers that are attended with Head Aches, Dizziness, Watchings, Convulsions, &c. It entirely removes all old Pains and Aches being applied to the Part affected. Spread on Leather.

O — Is a good Plaster to dress Blisters withal, and small Cuts (but not in the Legs lest it draw the Humors there) and for Boils, which it ripens, breaks and heals.

P — Is a good Plaster to heal and dry any ordinary Cut or Wound, especially in the Legs. Spread on Linnen.

Q — Dries and heals wonderfully; has all the Virtues of the former, and is excellent in all Bruises, especially where the Skin is broke. Spread on Linnen.

R — Is an extraordinary good Plaster, spread on Leather, and apply'd to the Part affected. It's good for old Achs and Pains (especially at the Stomach) Strains and Bruises, broken and dislocated Bones, and to strengthen weak Joints.

Allum, Tow and Leather, whose Uses are all too well known to want any Directions here.

In order to understand the due Application of the aforesaid Remedies, take the following Rules.

First, If any of your People should be visited with a Boil, or Swelling of any Kind, apply (A) 'till the Boil or Tumor is broke, then use (B) as is directed, 'till the Flesh is even with the Skin; then heal with the Pot (C) or the Plaster (P.)

Secondly, If any of your People should cut themselves at any Time, and the Wound bleeds apply (K) for the first Dressing; then use (B) 'till the Flesh in the Wound is even with Skin; then heal with the Pot (C) or the Plaster (P) [illegible]

Doctor's boxes contained every conceivable remedy, a welcome purchase for a long voyage.

Bark "REAPER" of Salem. Built at Amesbury in 1820.

Ship "ELIZA" of Salem

Altered to a Bark in 1838.

She made three whaling voyages between 1833 and 1841.

Merchant Princes of Salem's great age of sail, whose investments included whaling voyages to many world whaling grounds.

Dining room of the beautiful Pingree House and rear parlor. The quiet elegance of the interior and furnishings depict the gracious era when Salem held a prominent place in the maritime industries of New England.

KUNST VOOR KRACHT

VII.

SIet hier den grooten Walle-vis ,
Siet wat een monſter dat het is,
Siet hier dat onbelompen vee ,
Het ſchijnt een eylant in de zee,
En efter wort het dier gewont
Alleen maer door een gauwen vont,
Het wort gevangen en geſchent
Oock in ſijn eygen element ,
En dat noch van een ſchippers gaſt
Die nau acht voeten hoogh en waſt.
Vraeght iemant hier de reden van?
Wat is'et dat ick ſeggen kan ,
Als dat gewelt en groote kracht
Wort inder haeſt tot niet gebragt,
'tEn ſy dat wysheyt voor en nae,
'tEn ſy beleydt te roere ſtae.
Want anders ſiet men dat gewelt
Door eygen krachten wort gevelt.
Ghy bout dan niet op groote macht,
En min noch op uw eygen kracht;
Maer doet uw ſaecken met verſtant,
Want dat is vry al vaſter bant.

Master engraving from the original design of Adriaan Van de Venne, Jacob Cats, Amsterdam, Holland, J. Rotelband (1712). Appears in Volume 1 of 2 Volumes. The Dutch learned the art of deep-sea whaling from the early Basques. Courtesy N.Y. Public Library.

by the wind to the S and W. . . . a ship in sight on the weather beam.

Friday Nov. 3, 1837. Light breezes from S E to E at 10 A M spoke the *Ulysses Sherman* of Rio, 21 days from the Cape of Good Hope with 60 barrels of Sperm oil which he took of the River Plate; reports a severe gale having been experienced in the Indian Ocean. 3 whale ships foundered, several sail on shore at the Cape and at some part of Bombay, 14 sail dragged on shore in one night; the gale was experienced about September.

Wednesday, 29 Nov. 1837. Strong breezes from N E at 8 A M; saw whales struck one in the waist boat. The line parted, there was a large lot of them and during the day we lowered three times; got nothing.

Friday Dec. 8. Stiff winds from E N E steering N by W saw a ship steering S W. There is the same heavy sea here, but as yet we have seen but one school of whales. O favour us kind Providence!

Sat. 9. Moderate breezes N E with a heavy sea and no prospect of getting to the northward, saw nothing if I could only capture a whale, it would give my mind some little rest . . . I am now on the rack.

Sunday, Dec. 17. Gentle breezes from E N E, steering W N W, saw a humpback which we thought to be a Sperm whale, lowered for him, but soon found our mistake at 2 P M saw whale, took one to the starboard boat; took him alongside at 5 P M, and commenced cutting in.

At daylight commenced cutting again; got the head half off and saw the land under our lee about 20 miles distant; hove in the body and bailed the case, hooked on to the junk several times, but it was so rugged the hooks and straps tore out. After being satisfied that I could not save it without endangering the ship, I cut it adrift. When I took this whale, I was 99 miles from the land by the chronometer. I was satisfied the Chro was wrong, but did not think the error was so great. The rate when we left home was 2#, losing the correct rate is about 4# per day, the current set was 50 miles. Lat 25 15 S; Long. 47 48 S S W. It is now blowing heavy from E by N west. At the time we cut, it rained in torrents. When I saw the land the ship headed N, the land hove W N W to W by noon round heading S S E.

Sunday, Dec. 24. Saw whales, struck one to the W B Iron, drew, lost him.

Friday 29 Dec. Moderate breezes and pleasant from E; sent a boat on board of the French Barque *Hirondelle* 2 days from Fort Dauphin bound to Burbon, Capt. Teval, with a cargo of livestock, his Long. by dead reckoning 48.15 from Paris which would be 50.35 from Greenwich, mine 49.21 E by Chro. difference 0/1 1/14

Sat. 30th. Brisk breezes from E by N, steering W N W, until 2 M, then on the wind to the N. At 2 P M saw a large whale; struck him on the W B; the line parted; lost him; this is twice this line has parted, and we have lost two large whales by it. The lines were made for New Zealand, but by some unaccountable rascallity there is 20 yarns of New Zealand and 26 of tow . . . such is the faith to be put in Stickny. What can I do? This is my first voyage as master. It is expected by my owners that I should exert myself to the utmost to procure oil and I know that I have done all that man could do to find whales and have been successful thus far in seeing them for this I thank heaven, and for miserable lines I suppose I must thank Stickney.

Monday, Jan. 8, 1838. Very pleasant the wind from N E steering N by W; saw a large number of sulphur bottom whales and they turned flukes quite regular which gave us some trouble. At 2 P M saw school sperm whales, took one to the W B, at sunset took him alongside.

Tuesday. 9. Stiff breezes, commenced cutting; at 2 P M finished.

I have this whale's case inboard but should not care to do it again. At 4 A.M. move ship head S E. . . .

Wed. 10. Stiff breezes from S S E, headed and employed in boiling.

Thursday. Finished boiling.

Friday. Stowed down the oil in barrels. A sail in sight to the N W.

Saturday. 10 March 1838. Brisk breezes from E at 6½ A M saw whales and it is my hard lot to say that I lost my waist boat, she being stove into toothpicks by a whale. Mr. Boynton got injured in his leg and eye by falling through the boat, saved a small piece of her stern. Oh God forgive us and only crown our hard exertions with success.

Monday, November 12, 1838. Brisk breezes. Saw several whales. Lowered two boats in the forenoon. Could not fasten. The Barque *Octava* schooling round us. Did not lower in the afternoon. Killed two whales sunk them, so it goes. In four whales we have lost 22 irons, 1 lance (the whales got away with the weapons buried in their blubber). How long will it take to fill at that rate? Mr Boynton was taken out of the boat by the last whale. Fortitude preserved him. He got clear.

Dec. 17, 1838. Brisk breezes from N E at 9 A M saw whales and had the good fortune to take two. I am in a hell of a situation for men. Two of my men are sick. This leaves but 20. It takes 18 to man the boats, so that I dare not lower until one boat is fast. At sunset had one of the whales on board.

Tuesday, 26, 1839. Light breezes from N N E. At daylight put up for a whale in company with the *Martha* of Nantucket. The *Martha* struck a whale and lost him. Reports the *Bengal* in November with 400 whales, (bbls?) 40 Sperm, bound to New Holland. The *Endeavor* 150 of Sperm. The *Martha* lost four men by sickness at St. Mary's.

March 7, 1839. I have now been in this port 9 days. Most of the time bad weather with the wind N E, heavy showers of rain. I am now waiting for a wind or calm to go to sea. It is impossible to make sail from where the ship lays unless the wind is from S W or W, which is a rare thing at this season. In order to go out with safety the ship must be swooped out to the E N E or 1 or 2 miles.

Saturday, 9th. This morning at daylight hove short and carried out the kedge with three parts of whale line for the purpose of warping the ship out, but the wind blew so strong that we could not gain a fathom a head, let go the starboard anchor again and veered away the chains. At 9 A M went on shore and accompanied the governor on a fishing party. At 5 P M came on board with a heavy heart. If it is not calm soon, my case is a hard one.

Monday, the 11th. Last night slept on shore and let the girls come on board for the purpose of keeping the crew quiet. This is hard, but I am forced to do it. At daylight swooped the ship out nearly a cable's length, the sea beginning to make, could not possibly get her any farther and not having room to get underway was obliged to let go the anchors again. Stopped on board all day myself. At 1 P M sent Mr Boynton on shore with the boat; returned at 5. The crew wish to have the girls up tonight. I have told them it cannot be done. I have tried to reason the case with them but it is of no use. They have combined together tonight that they will do no more duty. Mr Hersey heard the agreement. It was this; that when they were called in the morning they would say that they were sick. The mates and Boatsteerers stand the watch tonight. The wind strong. (It seems they got a light wind at 4 A M and finally pulled out to sea. This on Wednesday, 13th of March, 1839. Eight men wouldn't do any work and had to be watched. They are headed from the Cape towards St. Helena).

May 26. Sunday. Saw the breakers on Nantucket Shoal.

May 27. Monday at daylight. Cape Cod in sight.

There is a list of stuff bought by the various men from the slop-chest such as trousers, socks, shirts, etc. also bread, beef, pork, coffee pot, tea, pepper, molasses.

Men listed for this voyage: James Jackson, Henry Williams, Sherwood, Weatherby, James York, William B. Cook, John Mitchell, John C. Very, Jr., Richard Bodwich, Johnson Laruck, George Richards, Francis Peach, James Moore, Mr. Hersey, Thomas James, Patrick Welch, Edward Page, Edward Mavehal, Benj. Neal, Benj. Wingate.

* * *

The log of the Ship *Izette* offers more detail of the actual chase and capture of whales. The *Izette* left Salem May 2, 1840 and returned in 1842. I learned that during World War II, Marton R. Lovett, who was employed by the W.P.A. Writers' Project, examined some of the whaling logs at the Essex Institute and found that the *Izette* notebook contained two separate logs, the Ship Izette and Bark Ann Parry. The Ann Parry, 348 tons, was from Portsmouth, N. H. 1824, registered June 20, 1849 in the names of Benj. West, John Hodges owners, W. M. Harron, master. She was not a Salem whaler. In rebinding the logs, no notice was taken of the Second Journal, detailing the voyage of the *Ann Parry*. Thus the two logs became almost inextricably mixed. Pages are sometimes upside down. Chronologically they are badly confused. The two distinct journals sometimes alternate. Both are incomplete and defaced. However, it appears that the logs of the *Izette* and *Ann Parry* were kept by the same log keeper, Augustus N. Perkins. He was 2d mate on the *Izette,* and 1st mate of the *Ann Parry*. From a reader's standpoint they are interesting and instructive. The style is concise and not too diluted with navigating detail. Both ships had a *greasy voyage*. Both were visited by death and accidents.

Ship Izette, Bay Whaling, New Holland, kept by Augustus N. Perkins, Lynn, Mass. 1840/1.

Friday. July 16. Blowing a gale of wind off the land. Four boats went down to the whale we sunk last night. Hauled him up and began to tow; the mate of the *America* fastened to a bull that ran faster than any whale I ever saw. He turned him up and four boats came and towed him to the ship. We towed our whale to the *Izette.*

July 17. Saturday. This morning the wind fresh. Cut in the whale and started the try works in the afternoon. The three captains took their boats and pulled up the Bay on an exploring expedition. Saw two whales, but did not get fast. Lowered 4 boats for a cow and calf but did not get. The captains got back at 7 P M. Trying out tonight. Winds light.

July 22. Thursday. The wind fresh this morning. Saw several whales. Lowered for a cow and calf. It came on to blow and we went on board again. This is dull music but do not be alarmed we should get a good voyage in the old ship yet, though it is fifteen months and over.

July 23. Friday. Moderate. Lowered for whales the Captain went in his boat. Could not get fast and went off on an expedition among the Islands. Got back at 5 P M. Captain Briggs got fast to a cow and calf. Killed and anchored her. The mate of the *America* got a Bull and took him along side of us. Winds light from seaward.

July 24. Saturday. Light winds. With rain at a strong breeze from the westward. Went down to look for the whale that we anchored, but could not find him. The mate of the *America* fastened to a whale but it came on to blow, and he sat her. Boyington killed a small bull and turned him to the *Barclay.* In the P M lowered the mate of the *America.* Got fast to a cow and got fast to the calf. Killed her and took her to the *America.* Ends light winds.

July 25. Sunday. Fine weather. The four mates of the *America* got fast to a cow and calf. Killed him and he sunk. Let him lay for the night. Mr. Ritchie got fast to a cow and calf. Killed the cow, but lost her, as she went off shooting blood. The mate of the *America* got fast to a cow and got stove. Mr. Boyington got fast, killed the whale and took her to the *Izette.* The two mates of the *Barclay* got fast and drawed. I got fast and cut the line with my lance. The mate of the *Barclay* got a small bull and took him to the *America* and both the *America* and *Izette* are trying out.

July 26. Monday. Light winds and pleasant. The mate of the *Barclay* and Mr. Boyington lowered for whales. Mr Boyington got fast. Killed the calf with the iron. Soon after fastened to the cow and his iron broke. At last the mate of the *America* got fast and turned her up and took her to the *America* at 9 P M.

Tuesday. 27. Light winds. Trying out. Mr Boyington and I lowered for whales. I got a cow and took her to the *America.* At 2 P M struck and cut with the lance a very large cow. The mate of the *Barclay* got a cow and the boats got her alongside ours as at ½ past 9 P M. Cooled down having tried out 40 barrels since morning. Mr. Ritchie has been aboard all day. Fine weather this night.

Tuesday Jan. 11, 1842. Begins with light winds and pleasant weather. Watch employed in job of ship duty. Cooper setting up shooks (for making barrels) at 4 P M got out the starboard boat. Middle and latter parts moderate. So ends this day.

February 20. Sunday. Begins with light winds and pleasant at 6 P M. Spoke Bark *Sophia* and *Eliza*—Coffin—Duxbury, 7 months, 30 barrels Sperm. Captain Coffin came on board at 9 returned. Throughout the night rainey; in the morning pleasant. Captain Coffin came on board. Saw plenty of Fin Back Whales. So ends this day. Latitude 22-25, Longitude . . . .

1842, March 8. Tuesday begins with fine breezes and pleasant. Employed in fitting rigen. Throughout the night pleasant. In the morning the same. The *Leonados* in sight. Bent over top gallant. Sails fore and aft. So ends. Land now in sight.

March 9. Wednesday. Begins with fine fresh breezes from S S E and pleasant. Watch employed in job of ship setting. At 4 P M spoke *Leonados,* Capt. Stewart came on board; at 8 returned. Middle and latter parts pleasant. So ends this day.

Thursday. Mar. 10. Begins with strong breezes from the south and pleasant. At 4 P M spoke Ship *Martha* and *Hudson.* Eight months, 4 Sperm Whales. Capt. Wellder and Stewart came on board. At 9 returned. Throughout the night pleasant.

March 11. Friday. Begins with fine breezes and pleasant. Sandy Island in sight. Came to anchor at 12 m. in 9 fathoms. The ship is loaded with natives. At 10 A.M. Moses Sargent died.

August 31. Coast of New Holland. Fine weather. Mr Ritchie got fast to a calf and broke his iron. I soon after fastened to the calf. Killed the cow. She sunk and we anchored her. Took the whale we anchored last night alongside of us and cut her in.

August 22. Saturday. Pleasant weather. Mr Boyington and Mr Ritchie towed up a whale; cut in the whale and began to *try out.* The *Garland* and the *America* got a whale each today and the *Barclay* went up to Tophams Cove to anchor. I got fast to a cow and soon had her killed. Two other boats came down and helped tow her to the ship. The *America* sailed on Monday.

August 24. Fine weather. The *Garland* got a whale today. I also got a large cow alone which is the third one I have gotten by myself. Mr. Boyington got fast to a cow and calf. She ran him about seven miles along the land. Mr. Ritchie chased her. At last the whale turned up on the breakers and drifted ashore. They hedged her out, towed her two miles and anchored her and came on board. The *America* got a whale. The *Barclay*'s got loose and was lost two or three times. Finally got a whale. It turned up just at night.

The captain of the *Emerald* decreed they would take both Sperm and Right Whales. The *Reaper* and *Izette* likewise took Sperm Whales. All three logs give glimpses of the exciting captures executed by Salem whaling teams. Feats of strength and daring were performed by officers and crews from their light rowboats with only hand harpoons and lances as weapons against the formidable foe, whom they usually termed an "ugly customer."

"The Cacholot (Sperm Whale) is not only better armed than the True Whale (Greenland or Right Whale) in possessing a great weapon at either extremity of its body, but also more frequently displays a disposition to employ these weapons offensively and in a manner at once so artful, bold and mischievous as to lead to its being regarded as the most dangerous to attack of all the known species of the whale tribe" (Frederick Debell Bennett, *Whaling Voyage Round the Globe,* 1840)

The oil of the Sperm Whale was preferable to any of the whales sought. Sperms, however, were ferocious fighters and some captains did not seek them out.

Here is a list of the whaling logs I located at the Essex Institute in Salem, Massachusetts:

Ship *Bengal* 1833-1835. To South Pacific. Kept by William Silver of Salem. Gift of William S. Neal, March 30, 1917.

Ship *Bengal* 1837-1838. Contains lists of raising whales in the back and Sperm Whales caught by men named.

Ship *Eliza* 1838-1841. Capt. Radcliff, Master.

Ship *Elizabeth* 1836-1840. To Pacific Ocean. Owned by Hon. Stephen C. Phillips, commanded by Isaac G. Hedge from Salem. Excellent state of preservation. Kept by Jonathan P. Saunders. Contains list of officers and crew, sketches of land elevations, charts and whales.

Ship *Elizabeth* 1840-1844. From Salem towards the South Seas, casting off from Phillips Wharf in Salem. Log contains account of sperm oil stowed down on board during stay in Pacific.

Ship *Elizabeth* 1844-1848. Worthin Hall, Master. To Pacific Ocean or elsewhere on a whaling voyage. Legible, well preserved log. More detailed treatment than usual whaling record.

Barque *Emerald* 1838-1840. To South Atlantic, Indian and Pacific Oceans. Joseph Dexter, Master. Log kept by William H. Nichols. Lists officers and crew, oil secured. Minimum of navigating detail. Much material on life, conditions, natural history, philosophy. This is perhaps the most outstanding journal of the group. Gift of William Henry Nichols, son of the whaleman, in 1913.

Barque *Emerald* 1840-1843. From Salem to South Seas. Commanded by Robert Brown. Log kept by John Gunny.

Barque *Emerald* 1843-1845. Voyage to Indian Ocean and elsewhere. Commanded by Edward F. Lakeman. Log kept by William R. Stanton. Separate home-made notebook gives account of provisions on board.

Barque *Henry* 1845-1847. Whaling voyage to South Pacific Ocean. Commanded by Charles Lind.

Ship *Izette* 1831-18- . Delagoa Bay. Commanded by Capt. Hart. Log kept by Augustus Allen, 29 years old. Gift of George H. Allen before 1930.

Ship *Izette* 1840-1842. This consists of two separate logs, *Izette* and *Ann Parry* inextricably confused and mixed. Both journals incomplete. Both kept by Augustus N. Perkins, 2nd mate on *Izette* and 1st mate of the *Ann Parry*. Interesting, instructive and concise style.

Ship *Lydia* 1835-1837. To Indian Ocean. Commanded by Capt. Ramsdell. Log kept by Abraham Hoxie. Good condition, hard covers. The writing is not too legible in spots. When the log ends with the *Lydia* along side of the wharf in Salem, there appears this apology initialed with two very fancy flourishes: "Excuse all bad writing and poor spelling." Log presented to Essex Institute by William S. Neal, Mar. 30, 1917.

Barque *Malay* 1839-1842. To Indian Ocean. Edwin Barnard, Master.

Schooner *Para* 1865-1867. Commanded by Oliver C. Hussey.

Barque *Palestine* 1839-1841. To South Seas. John Crimblish, Master. Log ends with these words: "There is talk of war with England and America—so ends this book."

Barque *Palestine* 1839-1841. To Indian Ocean. This appears to be another journal of the same voyage. Contains facts on barrels of oil secured, profits and losses. End pages missing.

Ship *Reaper* 1835-1837. This is probably the barque *Reaper* here called a *ship*. To Cape of Good Hope and return. This record is water-stained, fragile, with many pages torn and loose. In the back in a different handwriting are remedies for Hydrophobia, Rheumatism, corns, dropsy, snake bite and cancer. (For Cancer: Potash made of the ashes of . . . oak, boiled to the consistency of molasses applied as a poultice covering the whole with a coat of tar. 2 or 3 applications will remove all protuberances. After which heal the wound with common salve).

Barque *Reaper* 1837-1839. To Indian Ocean. Left Derby Wharf August 21, '37. Back of book contains an account of each whaleman's standing with the *Reaper*'s slopchest. Gift of William S. Neal, March 30, 1917.

Ship *Sapphire* 1836-1839. Journal kept by James Ward preceded by eight pages of Plane Sailing and nautical calculations.

Ship *Sapphire* 1839-1841. To Pacific Ocean. Commanded by Alexander Cartwright. Log kept by John Crowell. Predominantly nautical.

Ship *Sapphire* 1841-1842. Owned by Stephen C. Phillips. Capt. Alexander Cartwright.

Whaling Voyage 1836. Gift of N.D.C. Hodges, 1924. Original cover lost. No vessel named, but appears to apply to Salem.

CHAPTER

17

Salem's Greatest Whaling Year

"Success is full of promise till men get it, and then it is as a last year's nest, from which the bird has flown."
Henry Ward Beecher (1813-87)

Whaling as an industry was economically important aside from its adventurous phases. For the quarter century from 1835 to 1860, the whaling fleet of the United States averaged some 620 vessels annually with imports of 117,950 barrels of sperm, 215,913 barrels of oil and 2,323,512 pounds of whalebone. This represented a business of $8 millions a year, a substantial amount allowing for the change of values, which added much to the productive resources of this country. The greatest sperm year of all American whaling was 1837. In that year, 5,319,138 gallons of sperm oil were imported, valued at 82½¢ per gallon, or almost $5 millions.

In 1837 all of Salem's whalers were absent on long cruises. There were 432 Salem men employed in the whaling industry. From April 1, 1836 to April 1, 1837, this port was active with eighteen whalers, with an aggregate of 4,947 tonnage. This involved a capital outlay of $453,944. There were secured 108,065 gallons of oil valued at $40,866 and whalebone at 37,067 pounds valued at $7,535. Whaling interests of the town were further encouraged by the landing of a rich cargo of oil valued at $124,440.

In this hopeful year of Salem whaling, the subject was

brought to popular attention by the newspapers and quite effectively by the thoroughly-prepared lectures of the Rev. Charles W. Upham. The good minister added drama and interest by claiming Salem had whaled before famed Nantucket and with some assurance and local pride the *Salem Register* announced that the "outlook is definitely optimistic." At this time Salem was active as well with its regular fisheries. In 1837 fourteen fishing vessels worked out of Salem harbor and the codfish secured was valued at $16,552 and mackerel was valued at $21,450.

Whaling offered hope of remuneration and Salem merchants invested quite heavily. There was John B. Osgood who had much faith and did quite well. In 1837 he sent out six whalers: The *Bengal* under Captain Jackson to the South American and Indian Oceans which returned in 1840 hailing 1,800 barrels of sperm oil; the Ship *Izette* under Captain Hall to the Atlantic which returned *clean,* having sold out all its cargo; the Ship *James Maury* under Captain Bigelow to the South Atlantic and Indian Oceans which returned at the end of 1839 with 250 barrels of sperm and 2,050 of whale-oil; the Ship *Lydia* under Captain William Ramsdell to the South Atlantic and Pacific Oceans which returned before Christmas of 1840 with 400 barrels of sperm and 2,600 barrels of whale-oil; the Schooner *Mac* under Captain Emmons to the Indian Ocean which came back in the spring of 1839 with 500 barrels of sperm and 1,000 barrels of whale-oil; and the Bark *Reaper* under Captain Neal to the Atlantic Ocean on a short jaunt from May 24th to November 17th, returning with 70 barrels of sperm.

James King was responsible for the Bark *Cavalier* under Captain Francis to the South Atlantic and Indian Oceans in the fall of 1837 and on May 29, 1840 the *Cavalier* reported with 1,800 barrels of sperm. Mr. King also dispatched the Schooner *Franklin* under Captain Tracy to the Atlantic on September 5th for a ten months' voyage which yielded 110 barrels of sperm. The Bark *Derby* sailed for the South Atlantic and Indian Oceans under Captain Radcliff in July, 1837 and returned on March 24, 1839 with 180 barrels of sperm, 820 barrels of whale-oil and 8,000 pounds of whalebone.

John B. Pierce sent out the Brig *Emeline* to the Indian Ocean on July 15, 1837. The *Emeline* met with disaster at the Falkland Islands on April 15th of the following year, but shipped her oil home. Stephen C. Phillips' Bark *Malay* sailed under Captain Barnard for the South Atlantic and Indian Oceans on December 20th and returned after two years with 380 barrels of sperm and 1,000 barrels of whale-oil.

Joseph Hodges dispatched the Bark *Richard* under Captain Dewing to the South Atlantic and Pacific Oceans on August 21st. On May 28, 1839, the *Richard* returned with 500 barrels of sperm and 10 barrels of whale-oil.

Within the year 1837 can be counted some 21 whalers from Salem engaged in the fishery, a figure which includes the *Derby* which was wrecked in 1838 on the Falkland Islands. Starbuck lists 12 whalers registered during the year 1837. Additionally, there was the Brig *Eagle* which sailed as a tender or assistant to the Bark *Richard*. A look at the record of all American sailings of whaling vessels in 1837 sheds some light on the place held by Salem during its peak year of participation.

A total of 229 vessels left 38 American ports in 1837 on whaling voyages. New Bedford led the fleet with 61 sailings, about five times greater than Salem's figure. Sag Harbor, New York, appears second on the list with 24 sailings, twice as many as Salem sent out. Nantucket came third with 21 sailings. New London, Connecticut, had 17 sailings and Fairhaven had 14. Thus it appears that Salem was in sixth place with her recorded sailings of 12 vessels.

The rest of the ports sent out whalers in this fashion: Warren, Rhode Island, sent out 9 whalers and Bristol of that state sent out 7. Stonington, Connecticut, and Greenport, New York, each sent out 5 whalers, and Fall River, Massachusetts registered 4. Each of these ports sent out 3 whalers—Falmouth, Lynn, Dartmouth, Plymouth and Westport, Massachusetts; Hudson and New York City, New York; Newport, Rhode Island, and Bridgeport, Connecticut. The following ports sent out 2 whalers—Holmes' Hole, Provincetown, Newburyport and Rochester, Massachusetts; Wilmington, Delaware, and Mystic, Connecticut.

Represented by a single whaler each were Edgartown, Dartmouth and Boston, Massachusetts; Cold Spring, Poughkeepsie and Newburgh, New York; Providence, Rhode Island; East Haddam, Connecticut; Newark, New Jersey; and Portsmouth, New Hampshire. Aside from New Bedford, the rest of the American ports sent out 88 vessels and of these an even dozen came from Salem. The balance of 76 whalers were made up by the combined efforts of 36 seaports.

And yet Salem whaling was never to become truly successful. Its people were seafaring, its merchants progressive, and it was a natural consequence that this seaport should participate in a maritime pursuit, but its other interests took precedence.

CHAPTER

18

Shipwrecks, Disasters, Explorations

"Misery acquaints a man with strange bedfellows."
William Shakespeare

Just as Salem was getting a good grip on the whale fishery in 1837, a general business panic swept the country. It became the greatest financial setback experienced by the United States up to that time. Throughout the nation 618 banks were declared bankrupt and the blame popularly laid at the door of President Martin Van Buren. At New Bedford and other whaling ports, vessels lay idle at the wharves and Salem's ventures in whaling began to fail. Nevertheless, 1838 saw four whalers registered out of Salem: James W. Cheever's Bark *Eliza* sailed late in November and after an absence of three years, the *Eliza* gave a good account of herself, returning with 1,300 barrels of whale-oil and 275 barrels of sperm. Stephen C. Phillips invested in the voyage of the *Emerald* under Captain Dexter and John B. Osgood sent out the *Mount Wollaston* and the *Statesman*. The combined cargoes of these last mentioned two whalers accounted for 2,400 barrels of sperm and 1,100 barrels of whale-oil, valued at about $82,000 plus the value of the whalebone.

In 1839 Stephen C. Phillips registered the *Malay* which produced 1,300 barrels of sperm and the *Sapphire* which produced 800 barrels of sperm and 1,200 barrels of whale-oil. John B. Osgood's Bark *Reaper* imported 1,000 barrels of sperm. His

Samuel Wright of which much good was expected was lost at New Holland and never returned. Nathaniel Weston registered the Bark *Palestine* in May 1839 and she secured 1,700 barrels of sperm in the Indian Ocean. On October 16, 1842 the *Palestine* returned to Salem bereft of Captain Crimblish and her First Mate, George Coffin, both of whom died during the voyage in 1841. Built in 1835, the *Palestine* herself was wrecked after this unfortunate voyage.

Messrs. Phillips and Osgood were active in 1840 with four whalers. Mr. Phillips dispatched the Bark *Emerald* which returned with a cargo of 8,800 pounds of whalebone and 1,100 barrels of whale-oil. Mr. Osgood was responsible for the *Bengal* which imported 1,800 barrels of sperm oil; the Ship *Izette* which imported 900 barrels of sperm and 1,100 of whale-oil, as well as the *Mount Wollaston* which brought in 400 barrels of sperm, 700 barrels of whale-oil and 5,600 pounds of whalebone, having sold 1,200 barrels of whale-oil at Rio Janeiro.

In 1841 Mr. Osgood was represented by the Ship *James Maury,* which reported after four years with a cargo of 1,400 barrels of sperm, 500 barrels of whale-oil and 3,600 pounds of whalebone, a good yield for which Captain Benjamin R. Hussey and First Mate Charles F. Pinkham both paid with their lives. The captain died June 15, 1844 and hardly three months later was followed by the First Mate. After many hardships, the *James Maury* finally arrived in Salem in mid-winter of 1845.

Stephen C. Phillips sent his *Elizabeth* to the Pacific in January of 1841 and she returned in October of 1844 with 1,500 barrels of sperm. James W. Cheever's Bark *Eliza* secured 200 barrels of sperm which she sent home as she was condemned as unseaworthy at Tahiti in July of 1843. The Bark *Henry* under Captain Manchester went to the Indian Ocean in the fall and returned in April, 1845 with 2,400 pounds of whalebone, 140 barrels of sperm and 300 barrels of whale-oil.

In 1842 there were only two whalers out of Salem. Mr. Phillips registered his *Malay* under Captain Lakeman and Mr. Osgood registered his *Statesman* under Captain Elisha Doane. The voyage of the latter vessel produced 500 barrels of sperm oil. Both vessels met with misfortune. The *Malay* was lost in shipwreck and the *Statesman* had suffered so much battering by storms that she was condemned as unseaworthy. The next year, 1843, there was only one vessel registered from Salem. This was the *Emerald,* which never returned to the home port, as she, too, met with disaster.

In 1844, Mr. Phillips sent out the *Elizabeth* under Captain Hall just a week before Christmas. The *Elizabeth* took the long trek around Cape Horn to the Pacific. The voyage was successful and on May 15, 1848, the *Elizabeth* returned with a cargo of 620 barrels of sperm, 1,580 of oil, and 14,000 pounds of whalebone, worth about $35,620. In 1845, Mr. Cheever sent out the whaler *Henry* to the Pacific under Captain Lind. She was wrecked at the Marquesas Islands. 800 barrels of whale-oil were taken from the *Henry* before she sank.

With so many factors in the favor of making Salem a successful whaling port, why did it fail to do as well as the other whaling centers? As we have seen, adversity played a large part. At least eleven whalers met with destruction which brought great losses to their owners. The *Catherine* burned off Oahu, Hawaii, in 1834, a disaster usually ascribed to the *Charles Doggett,* a brig which was lost on Oro Touga in the Fiji Islands in 1836. (Starbuck and *Salem Ship Register,* p. 205). The *Derby* crashed on the rocks at the Falkland Islands in 1837 and the *Emeline* was lost on the same islands on April 5, 1838. The *Palestine,* as noted, was wrecked four years later. On July 8, 1840 the *Samuel Wright*'s hull was fractured and she went to the bottom of Geography Bay, New Holland. It was impossible to save the 450 barrels of whale-oil on board which sank with her.

In 1843, the Ship *Sapphire* foundered at sea. On July 27, 1842 the *Malay* was wrecked on Europa Rocks in Mozambique Channel on the east coast of Africa. On June 15, 1843, the Bark *Eliza* loaded with sperm, was reported at Tahiti with a bad leak and was subsequently declared unseaworthy. The roll call of disasters continues. On November 3, 1844 the Bark *Statesman* was condemned at Talcahuano, a seaport of southern Chile. In 1845, the Bark *Henry* met her fate in a heavy storm off the Marquesas Islands, one of the Polynesian group in the Pacific, and what was salvaged of the wreck was sold at Tahiti. In the same year, the Ship *Emerald,* which had so nobly served the whaling industry, met her end off Fort Dauphin, Madagascar, on March 18, 1845, but her complete cargo of oil was saved.

There was insufficient capital available to overcome such losses. The decline of Salem whaling soon followed the series of disasters to the Salem whaling fleet. In the absence of replacing the lost ships, no real recovery of Salem's whaling industry was possible. By 1845 there were only four whalers active from Salem and 110 men employed compared with 432 men employed in 1836. This was a disheartening outcome in view of the progress being made in the industry as a whole. In 1842, reports had shown that America owned about three-fourths of the total whaling tonnage of the world. The number of ships had almost doubled between 1832 and 1842 and the year 1844 found the United States the owner of 644 whaling vessels.

America surged forward to its greatest whaling year of all times. In 1846 there was a fleet of 736 vessels flying the Stars and Stripes. The whalers were worth $21 million and there was an investment in the industry itself of $70 million. It is estimated that 70,000 persons were employed as whaling became the most adventurous and flourishing of American industries.

The cause of so much progress was the whale himself. One must recall that the whaling boom of 1846-7 coincided with the opening of new whaling grounds for Bowhead Whales in the seas of Okhotsk and Kamchatka, the inlet of the North Pacific Ocean on the East Coast of Siberia. In 1847 the American fleet comprised 727 vessels from 34 ports. New Bedford was active with 254 whalers. Nantucket had 75. New London, 70. Sag

Harbor, New York, had 62. Fairhaven registered 48. Stonington, 27. Warren, Rhode Island, 23 and Mystic, Connecticut, 17.

Thus eight ports whaled extensively and twenty-six smaller ports contributed the balance of America's whalers. After the ruthless war waged on the whale during the boom, the whale herds were seriously reduced. A decline set in during which many of the smaller ports were forced out of the field. This development included Salem which did not register a whaler between 1845 and 1851. Then it became active again with a single entry almost yearly until after the Civil War.

After the decline, there was a revival of whaling due to the opening of the Arctic fishery by an American vessel. In 1848 the whaler *Superior* from Sag Harbor, New York, commanded by Captain Royce, pierced the mystery of Bering Strait and passed into the Arctic Ocean, that most northerly part of the Pacific which separates Asia from Alaska. Fog and ice make the region impassable except for the period from November to May. The *Superior* became the first ship to enter the hitherto unnavigated waters of the Arctic Ocean and was rewarded by finding Bowhead Whales in abundance. This frozen area was the last whaling ground opened up by the American whalers who had served as explorers and advance guard into so many unknown places of the globe.

The story of American whaling captains as explorers and discoverers has not yet been completely told. As early as 1819 the Russian explorer Von Bellinghausen was surprised to find American whalers using as a base of operations the yet uncharted islands of South Georgia in the Antarctic. Twenty years before the Russian's visit, the Salem ships *Minerva* and *Rachel* were sealing in the Antarctic. The complaints to Congress by Nantucket whaling merchants in 1828 and 1834 that their vessels were being shipwrecked on uncharted islands—as many as 150 of them unregistered on any map, led to the expedition in which Commander Charles Wilkes, in command of what was probably the most important hydrographic survey since Captain Cook, spent four years (1838 to 1842) in a globe-circling expedition under the auspices of the United States Navy, during which he discovered the Antarctic continent.

Between the close of our Revolution and our Civil War, Americans were at home in every part of the Pacific, once a vast whale fishery. By 1840 the smoke from the try-works of hundreds of American whalers blackened the skies with 15,000 seamen manning the American whaling fleet. The groundwork had been carefully laid for this success. Following the War of 1812, bold traders from Salem, Massachusetts, Stonington, Connecticut, and many New England ports dropped anchor at New Guinea, the Solomons and the Admiralties. Whale-oil, sealskins, bêche-de-mer, edible birds' nests, turtle shells, sandalwood and sea bird guano, a valuable fertilizer high in phosphates, made up the unusual cargoes which sold at fabulous profits in Canton, China, San Francisco, New York, Boston and the ports of Europe.

American discoverers in the Pacific area were many. Most of the adventurers were whalemen or connected with whaling and trade.

The complete story of discovery by whalers is yet to be told, but we do know sufficient to show the impact upon the Pacific area by the old whalers out of New England. Capt. Joshua Coffin of the *Ganges* out of Nantucket discovered Gardner Island in the Phoenix group on a voyage between 1821-27. Captain James J. Coffin of Nantucket in command of the British Whaleship *Transit* is credited with the recognized discovery of the Bonins, or Ogasawara Shoto, stretching northward between Marianas and the Japanese coast. He called the most southerly cluster he found the Coffin group. Four of them he named Fisher, Kidd, South and Pigeon Islands. American whalers were acquainted with the bleak Izus, north of the Bonins, but like the Volcanoe Islands which include Iwo Jima, found them useless for refueling or resupplying their ships.

Whalers contributed much to the discovery of islands within

the Hawaiian group. The first American whaler to reach the Pacific is said to be the *Beaver,* Capt. Paul Worth, sailing from Nantucket in August, 1791.

In 1820, the year after the *Equator* and *Balaena* had reached Hawaii as the first whalers, Captain Joseph Allen in the *Maro* of Nantucket charted Gardner's Pinnacle and Maro Reef, while Capt. Gardner in his *Balaena* discovered Wake in 1823. Its position, however, was not definitely fixed until 1841 by Lt. Wilkes.

A chance meeting unveiled a now interesting section of the Pacific. Captain Allen ran into his old friend Capt. Jonathan Winship of Salem, Massachusetts, who told Allen of great schools of Sperm Whales particularly dense in the region northeast of the Ladrones (Marianas) which were first discovered by Magellan in 1521, some 1300 miles west of Wake. Using this tip, Captain Allen in the *Maro,* in company with another Nantucket Captain Frederick Coffin in his Ship *Syren,* sailing a north west course from Honolulu discovered the famous Japanese grounds of the whalers. Guam, Saipan and Tinian became familiar stopping places of American whalers.

Some early explorers probably sighted the Marshalls and stopped there, but the Marshalls and Gilberts were named for the British sea captains who discovered several islands in each group in 1788. This area of the Pacific was cruised by American whalers early in the nineteenth century. Captain George Ray in the *Nantucket* discovered Ebon Atoll southernmost of the Marshalls in 1824, naming it Boston Island. It is possible that Capt. William Plaskett of the *Independence* out of Nantucket discovered Nonauti Island in the Gilberts in 1822.

In the British-controlled Ellice group which stretch about 15° north to below the equator, at least four islands are claimed to have been discovered by Nantucket whalers to which the old files of the Nantucket *Inquirer* and *Mirror* testify. Captain Obed Starbuck of Nantucket has been credited with the discovery of Niutao which he named Loper Island after his ship. Captain George Barrett in the *Independence 2d* of Nantucket discovered in 1821 the two southernmost Ellice Islands, Nurakita and Nukulaelae. He named the first Independence Island and the second Mitchell Island for Aaron Mitchell the owner of the vessel.

Following the War of 1812, traders from Salem, Massachusetts, Stonington, Connecticut and many New England ports dropped anchor at New Guinea, the Solomons and the Admiralties to take whales and seals for profit in Canton, China, San Francisco, New York and Boston. Most of the adventurers in these exotic islands were whalemen or connected with whaling and trade. Nathaniel Savory of Massachusetts settled in Chichi Jima off Japan in 1830 and prospered in the rum trade with the whalers. Long before Commodore Perry opened Japan to intercourse with the rest of the world, the Sea of Japan had been visited by American whalers. Captain Riketson in the *Hercules* and Capt. Turner in the *Janus* of New Bedford took whales in 1843 off Kamchatka Peninsula, Sea of Okhotsk.

As early as 1835 whalers pursued their quarry in the Northern Pacific. Capt. Barzillai T. Folger sailed the *Ganges* to the Kodiak grounds in that year and became the first whaler to operate off the Aleutians. In 1848 the Whaler *Superior,* Captain Royce from Sag Harbor, New York, passed through Bering Strait, most northerly part of the Pacific into the Arctic Ocean which separates Asia from Alaska. The hitherto unnavigated waters of the Arctic Ocean became the last whaling grounds opened up by American Whalers.

It is not unusual that the whalers should be responsible for discoveries. By 1840 there were at least 675 American whaling ships in the Pacific. By the time of the Civil War, with the decline of our merchant marine, the replacement of whale-oil by petroleum, and the concentration of the development of the American West, interest was distracted from the Pacific and its mysterious islands. When permanent military bases in the

Pacific were recommended to the Senate Naval Affairs Committee, all the islands proposed were known to American explorers, seamen, missionaries, scientists and whalemen of more than a century ago with the exception of Midway discovered by Capt. N. C. Brooks and claimed for the United States in 1859.

The United States Navy began World War II using nineteenth-century sea charts. Battles had to be terminated because nobody knew for certain where the bottom of the sea was at given places. As late as 1941 entire archipelagoes in the Pacific were of interest only to oil and soap companies. On Guadalcanal, the first land engagement was waged on the Ilu River, but participants thought it was on the Tenaru. Our century was unfamiliar with the Pacific. Some islands were uninhabitable. Army engineers succumbed to cerebral malaria and battles were waged under incredible conditions. Shipwrecked sailors were devoured by sharks. In this hard war ruthlessness became the keynote.

With only a toehold on New Guinea and Guadalcanal, by late July 1941, Washington, D.C. froze Japanese assets and cut them off from their oil supply. December 7, 1941 brought Pearl Harbor with the loss of the American fleet. Japanese troops poured into Indo-China. Since nobody believed Japan would strike, Congress refused to fortify Guam. Admiral William F. Halsey, Jr. believed Japan would be defeated by 1943, but the Japanese followed up Pearl Harbor with telling blows, launching simultaneous invasion assaults on Hong Kong, Malaya, the Philippines, Guam and Borneo. The second World War again made the names of Saipan, Iwo Jima, Okinawa, Guadalcanal, Samoa and Wake household words.

Familiarity came with sacrifice. It came with Black Sunday, the day which lives in infamy, when 2,300 American sailors, soldiers and marines perished in a few nightmare hours of attack. In the battleship *Arizona* alone, 1,102 men were drowned or burned to death when a bomb exploded her ammunition magazines and another bomb by some fluke went down a stack. The cost to the United States Navy was more men than it lost in the Spanish-American War, World War I and the Korean War combined. It was achieved by Imperial Japan's striking force of 6 aircraft carriers, 9 destroyers, 2 battleships, 2 cruisers and 3 submarines, launching out from Tankan Bay in the Kuril Islands.

General Hideki Tojo's military government moved to fulfill Japan's destiny. To support this formidable sea power were waves of 40 torpedo bombers, 50 high-level bombers, 50 dive bombers and 40 fighters, emblazoned with the Japanese ball insignia which roared toward the northerly coast lines of Oahu. Their low-level bombers bracketed the island from the west across Waianae, from the east over Diamond Head, fighters bored inland strafing American airfields as Pearl Harbor met its doom. The battleships tied up along Ford Island were the hardest hit.

The United States Pacific fleet lost: *Arizona, Nevada, Tennessee, West Virginia, Maryland, Oklahoma and California.* The *Arizona* suffered most of all. On the signal bridge Rear Adm. Isaac C. Kidd was killed. With a cracked keel, the *Arizona* sank in a convulsion of smoke and fire to become the tomb of 1,102 men. A part of the ship still protrudes as a reminder of the Great Unpreparedness in the Pacific. The *Arizona,* incidentally, is still commissioned. In a rare departure of sentiment, the United States Navy officially regards her as a member of the fleet, because her crew is still on board. In the wake of Pearl Harbor, all other disasters in this area from the sailing days forward, pale in comparison.

CHAPTER

19

The California Gold Rush

"Such stuff the world is made of."
William Cowper (1731-1800)

In 1849 the California Gold Rush almost ruined the prospects of American whaling. It was customary for Pacific whalers to refit at a Pacific port, to discharge oil for shipment home by commercial carriers and to take on men and supplies. The discovery of gold and its consequences disrupted this routine. Ironically it came the year after Captain Royce in the *Superior* opened up the Arctic Ocean theater for whaling operations. The rush for California's gold threatened to destroy Pacific whaling. Here was a madness, an obsession, an obliteration of all other ambitions of acquisition, appealing instantly to adventurers. Gold! gold! gold! In the glittering metal, the liquid gold of the try-pots was forgotten.

No town in New England was free from the infection of the gold fever which swept the nation. Definite information of the discovery had reached Salem in October of 1848 as the Brig *Mary and Ellen* was fitting out for sea. It was Stephen C. Phillips and some others who saw to it that suitable cargo was quickly put on board. Under Captain J. H. Eagleston the brig cleared for Hawaii via California on October 27, 1848 and became the first vessel to sail from Massachusetts for California after the gold discovery. On November 23, 1848 the Salem Bark *Eliza* (not to be confused with the Bark of the same name which ten years earlier went whaling for James W. Cheever) cast off from Derby Wharf bound for California. The *Eliza* is unforgettable in Gold Rush annals. It was her exuberant passengers who were the first to sing the verses which in a hundred different versions became the theme song of the Forty-Niners. To Stephen C. Foster's hit tune "Oh! Susanna" the men sang their crazy words:

> *"I came from Salem City*
> *With my washbowl on my knee*
> *I'm going to California*
> *The gold dust for to see*
> *It rained all night the day I left*
> *The weather it was dry*
> *The sun so hot I froze to death*
> *Oh! brothers don't you cry*
> *Oh! California! That's the land for me!*
> *I'm going to Sacramento*
> *With my washbowl on my knee.*
>
> *"I jumped aboard the* 'Liza *ship*
> *And traveled on the sea*
> *And every time I thought of home*
> *I wished it wasn't me*
> *Oh, California! etc. etc."*

All sorts of wild rumors found their way back to Salem and one Salem adventurer, who had gone early to California, wrote home that "In twenty days of digging I got $1500 worth of gold." Whaling profits could not hope to compete with the kind of adventure which in the twinkling of an eye could enrich a man for a lifetime. It was inevitable that the Gold Rush should adversely affect what remained of Salem's whaling industry after the disasters. There was the case of the Ship *Elizabeth* whose record as a whaler was an enviable one. The *Elizabeth* had narrowly escaped destruction at the Western Islands in the Azores in 1839 and rode through an earthquake at sea. In 1841 and 1844 this vessel had undertaken two arduous whaling voyages. From the second she returned on May 15, 1848. This proved to be the *Elizabeth*'s last voyage as a whaler. Pressed into service for the new El Dorado, the *Elizabeth*'s next long trek was to San Francisco in 1849 as one of the gold hunters. Among the names of her owners for this voyage to California appear those who had been long associated with whaling. No disaster is as final as the suspension of interest by an industry's underwriters.

Many ships and men on the New England coast converted to a new form of Mammon worship. Men armed themselves with picks, spades, the equivalent of "K" rations, and banjos, took their departure from family and friends with good wishes ringing in their ears. Transportation was at a premium since there were far more eager passengers than available ships to take them west. California lines of packets advertised and secured buyers for cabin passage at $250 and steerage passage at $150, recommending early purchase of tickets. But why spend such a fortune for transportation with the outcome itself a gamble? Those familiar with the sea knew they need pay nothing at all to undertake a whaling voyage to the Pacific. With the deliberate intention of reaching San Francisco to go gold mining, many men lied their way on to whalers departing from New England. Upon arrival in San Francisco, they deserted ship.

The practice spread until whole crews, including captain and officers, deserted. Why go any farther? A whale was a whale if and when you could catch up with him. The gold fields were *here,* beckoning with a glittering invitation. At last, shipping on a whaler came to be recognized as a sure means of reaching California from the eastern states. Whaling interests lost thousands of dollars as valuable equipment was abandoned and the object for which the vessel had been provisioned and outfitted was forsaken. The whalers themselves were left idle in the harbor. If a Captain and his officers remained faithful, it was impossible for them to secure enough men to continue the whaling cruise.

The Gold Rush year of 1849 found only fourteen vessels whaling under the British flag. The supremacy of Britain as a whaling power had long gone to American ships. By 1849 America had passed its zenith but still possessed a whaling fleet of 596 vessels manned by 18,000 seamen. In addition to the Gold Rush, there were other important forces which mitigated against the continued rise of American whaling. Solvent capital in New Bedford began to trickle into other channels. In 1846 the queen of American whaling ports began to manufacture cotton goods, an undertaking which was eventually to rise to first place in New Bedford's productivity. Salem, too, began to turn its attention from the sea to manufacturing as early as 1839 when the Naumkeag Mills began operations at Stage Point. A year before the *Eliza* made her celebrated voyage of 1848 to the gold state, the Naumkeag Steam Cotton Company had emerged in full operation from its experimental predecessor. As Salem ceased to be an outstanding seaport, shipping men and merchants found new investments for their capital and Salem's factory era was under way.

The New England of the whaling years saw every inlet and cove a thick forest of spars and masts of the white-winged wooden ships of sail. Gradually New England turned from seafaring to manufacturing and factory furnaces replaced the ovens of the whaling try-works. Perhaps it was timely, for the introduction of steamboats made many of the small harbors unsuitable for the navigation of heavier vessels with their deeper draw. In addition to economic changes, there was the challenge to strong men of raw America. Many of the adventurous were drawn to the development of pioneer lands and miles of prairie schooners took the long trail toward the Pacific Coast to open up the great West.

It was a time in America's coming of age when many epics were being written in factory smoke, on the seas, on the land and also in the expanding nation's literature. Two important books were published at mid-nineteenth century pertinent to this theme, one a product of the Salem heritage, the other indirectly connected with the old town. Nathaniel Hawthorne's *Scarlet Letter* appeared in 1850. It was produced in part during the author's tenureship at the Salem Custom House. The other book was Herman Melville's powerful saga of the whaling industry which appeared in 1851. It is odd that Fate has so long denied to Salem any recognition as a whaling port, yet prompted Herman Melville a century ago to dedicate his *Moby Dick or The Whale* to his Salem friend, Nathaniel Hawthorne.

It was Hawthorne's destiny to see the passing of Salem as a great seaport. The era of adventure, exploration and opening up exotic markets was about over when Hawthorne was appointed Surveyor of Customs in April of 1846. Salem was then a picturesque elm-embowered town hardly recognizable in the commercial city of today. Bearded sailors from distant lands moved about noisy Derby Street where the handsome Custom House of *The Scarlet Letter* still looks out on what is left of Derby Wharf. There were haunted houses and gabled roofs under which were the scanty old libraries and relics of the early Puritan colonists. There was a steady importation to Salem of sea-legends, tales of slavers and pirates. And of more recent vintage were the high adventures and heroism of the whaling voyages.

In Hawthorne's first year of office at the Custom House, some whalers were still absent on long voyages. There was the Ship *Elizabeth* for one. In 1842 the Whaler *Malay* had been wrecked in Mozambique Channel. In March of 1845 the *Emerald* had met her end at Madagascar, and the *Sapphire* had foundered in the West Indies. In 1844 the Bark *Statesman* had been condemned as unseaworthy. And all these records were reading material for the author whose father, Captain Daniel Hathorne, died of a tropical fever at Surinam during a prolonged voyage.

The Rev. Joseph Felt summarized the case for Salem whaling in a brief entry made for the year 1847 in his *Annals of Salem*. (Vol II p. 225)

"The prospect is that this perilous employment recommended in hope as to its increase, continuance and profit, will soon terminate in disappointment."

Whalers continued sporadically to leave Salem but not with

the old-time vigor of the 30's. In 1851, we find the Bark *Margaretta* registered by Benjamin Webb for a whaling voyage to the Atlantic under Captain Prior. In a year the *Margaretta* returned with 320 barrels of sperm and 37 of oil. The next year, the *Margaretta* sailed again under Captain Holmes and secured only 180 barrels of sperm, according to the customs report, and 3 barrels of whale-oil. As no whale yields so little of himself to the industry, it is obvious that Captain Holmes sold his cargo before reaching Salem. The year 1853 saw no registrations, but in 1854 Mr. Webb was again active with the Bark *Messenger* which went on a two-year cruise to the South Atlantic and returned with 362 barrels of sperm and 114 barrels of whale-oil. Returning in August of 1859, the *Messenger* was not sent out again.

CHAPTER

20

Whaling Goes West to San Francisco

"It is not only old and early impressions which deceive us; the charms of novelty have the same power."

Blaise Pascal (1623-62)

Few industries present a more interesting story than whaling. Few industries offer an opportunity for such a complete study of rise and decline. The whole history of whaling is repetitious. A species of whale is discovered. A maritime people devise ways and means of securing it. Great profits are made. The whale grows scarce. Eventually he disappears. The whaling industry dies.

Whaling is as much a part of America's robust past as the settlement of the Northwest Territory, the Gold Rush, the Pony Express, and the bison herds which roamed the western prairies. The rewards of whaling to the nation were many. In war, it gave to our navy some of its sturdiest and bravest seamen. In peace, it gave to our commercial marine its choicest and most skillful of officers. Today, whaling holds an important place in the economy of several nations, and only a hidden place in our own country.

Today's modern whaling industry seems strangely unrelated to the form it took centuries to develop to its triumphal role of the 1800's. Hardly any of the savor remains of the old whaling industry. It is more than an absence of the once familiar "Thar she blows!" with its Yankee twang. Science has removed the dangers and most of the thrills of whaling. Radio, aviation, electricity and radar are the twentieth century version of whaling as man continues his mastery of the sea.

New hunting grounds had to be found as the whales continued to retreat before continuous aggression. By 1761 the North Atlantic was practically devoid of Right Whales. The British and Dutch had moved to Davis Straits between Labrador and Greenland and soon the Greenland Right Whale was all but exterminated in his habitat. As we have seen, the hunt then turned to the Sperm Whale which frequented the warmer seas. Before 1800 British whalers had entered the Pacific. New England whalers soon followed them and by 1850 almost every ship in the American whaling fleet was in the Pacific. At the close of the Civil War, the Right Whale had practically been removed and the number of Sperms began to dwindle. Despite the inroads made by petroleum, whale-oil and whalebone were much in demand and brought high prices after the Civil War. Whaling soon enjoyed a revival.

As the industry declined on the Eastern seaboard, it tended to move to the Western coast. With the Atlantic fished out, the North Pacific and Arctic became the most frequented whaling grounds. Even there the Bowhead was growing more scarce after so many years of exploitation. The groundwork for a Western whaling port was laid as early as 1850 when the Whaler *Popmunnett* left San Francisco, for the Galapagos Islands. The quarry was the Sperm Whale. The first expedition from a coastal port in the West was made in the year California was admitted to statehood in the Union. By 1852 there were eight whalers registered out of San Francisco. Whaling was unsuccessfully attempted from Monterey Bay, 120 miles south by east of San Francisco where a colony of Portuguese and Italians had congregated.

An event of importance to the whaling industry was the purchase from Russia of Alaska in 1867. The Americans thus secured their own waters in which to hunt the elusive Bow-heads. The Alaskan purchase brought San Francisco closer than any port to the northern hunting grounds. By 1869 this boom town of the Gold Rush was well launched on its whaling career, as Nantucket dropped out and the transcontinental railroad replaced the Pony Express.

It was a long time before New England ceased to whale. In 1871, the year in which Salem's whaling came to a close, New Bedford sent out 41 whalers, Provincetown 10, New London 8 and the balance of vessels dispatched from the East were divided among the small ports. Boston and Marion, Massachusetts each sent out 2. Westport sent out 2. Fairhaven, Edgartown, Beverly, Massachusetts and Sag Harbor each sent out 1 whaler. San Francisco, too, had only one registration for 1871.

The Arctic whaling disasters of 1871 and 1876 proved a heavy blow to America's declining industry. Many still hoped to revive the pursuit to its old-time vigor, but the odds were too great. Alaska is now a state and it is not inappropriate to recall the part played by Yankee whalers there both before and after its purchase from Russia a century ago. The Yankee whalers were chief among the pioneers of the English language and American merchandising in the Russian Far East.

Yankee whaling began in the seas of Siberia as the Whaler *Superior* of Sag Harbor, Captain Royce, entered Bering Strait in 1848. There is no record of any whale hunts there before that date although Peter Dobell of Pennsylvania had been the first American to appear in Kamchatka as early as 1820. He wanted very much to bring Americans into Russia's Far East. By 1879 the Russians were sufficiently concerned about Yankee influence to send some cruisers to north eastern Siberia before that area "forgot it belonged to Russia." The "Yankee influence" consisted of the colorful seamen from New Bedford and Salem, from Sag Harbor and Cold Harbor who had come to catch and to try-out whales of the Sea of Okhotsk and other Siberian waters.

Perhaps no other kind of whaling offers such rich material as

Arctic, or Bowhead whaling, yet there is little of it on the records. The hunting of Bowhead Whales was prosecuted in a region beset with inconceivable hazards with the loss of many ships and lives in this remote icebound region. The natives were unaccustomed to keeping records, and little remains beyond surviving whaling logs to tell of this particular phase of Russian-American relations. Practically unknown to civilization, no writer ever visited it although American whalemen from 1848 went there every year for forty years. Yankee whaling served to draw Russia's attention to its Far East. It did not give rise to successful Russian whaling in spite of sporadic efforts, well into the Soviet years, to bring back the old profitable hunting.

The early discovery of Spitzbergen by the Dutch navigator, William Barentz, is recalled and his death in Nova Zembla in 1597. Part of his crew escaped to report vast schools of whales. This hastily brought to the scene whalemen from many European nations. The Greenland whaling which followed the discovery of Spitzbergen, with such tremendous profits, has long been extinct and *Smeerenburg,* or Blubbertown, built by the Dutch near the North Pole forgotten. A Norwegian explorer in 1871 came upon the Spitzbergen winter quarters constructed by William Barentz some 275 years earlier. An important find was part of this explorer's faded journal. Whaling circles in 1871 were as much stirred by this event as the man on the street who, at the same time, saw strange visions under the seas, on the moon and in the air, revealed by a new book called *Around the World in Eighty Days* by Jules Verne.

By 1877 the entire American whaling fleet was reduced to 163 vessels—112 of which were ships and barks and 51 brigs and schooners, chiefly out of New Bedford. By 1880 New York and Rhode Island ceased to be whaling states. Only Massachusetts and Connecticut kept up the whaling tradition. By 1886 it was mostly a spectral fleet which remained of the once mighty whaling Armada. After 1880 the growth of San Francisco as a whaling port was rapid. It had averaged only 8 vessels annually between 1869 and 1880. The opening of the transcontinental railway and the use of a combination sail and steam whaler enhanced San Francisco's whaling possibilities.

The rise of San Francisco signified the transfer of whaling interests from East to West. Technically, New Bedford possessed the larger fleet, but a great many of its vessels carried on the trade with San Francisco as headquarters which was true also of New London. For instance the *Charles W. Morgan* sailed from New Bedford from 1841 to 1886. On December 3, 1887 she began to sail under a San Francisco registration and continued to do so until June 12, 1906. After this date, the *Morgan* returned to New Bedford. During the time this whaler sailed for San Francisco, she was owned by the same New Bedford Company to whom she was re-transferred in 1906. This was J. & W. R. Wing.

Refitting and provisioning eastern ships in San Francisco helped a great deal. There still remained the difficulty of transferring to the eastern refineries the oil procured. In 1883 this obstacle was overcome when refineries for oil were built at San Francisco as well as factories for the making of sperm candles. In 1893 there were 33 whaling vessels registered at San Francisco. About 22 of them were steam propelled. For the ten-year period ending in 1905, the whaling fleet out of San Francisco averaged 37 vessels a year. The yield in whaling products was valued at one million dollars annually.

Fourteen whalers were active out of San Francisco in 1906, the year most of the seaport was reduced to ashes by earthquake and fire. At this time whalers were still regularly registered from Boston, New Bedford and Provincetown, the only ports in the East still harboring active whalers. Long after petroleum derivatives and vegetable oils displaced whale-oil, New Bedford continued to dispatch its whalers to distant oceans. The West Indian fishery and that of the Southern Indian Ocean which so many Salem whalers had hunted to their profit, were no longer followed by Americans. The American fishery in Davis Strait and Hudson Bay in the Arctic, comprised only a few vessels from 1890 to 1897. The only fishery remaining to this country of any magnitude was that from San Francisco which concentrated on the North Pacific and Arctic. Added to the oft-demonstrated danger of the ice packs to the wooden whalers, there was the great distance in reaching the whaling grounds. From San Francisco to the Diomede Islands in Bering Strait it is 2,860 miles. From there to the mouth of the Mackenzie, where the whales were comparatively abundant, it is another 870 miles.

CHAPTER

21

The Civil War and Whaling's End

"The end crowns all, and that old common arbitrator, time, will one day end it."

Shakespeare

Whaling pursuits continued as the nation moved closer to open conflict in its War between the States. The *Falcon* was idle from July 26th to August 31st, 1859, scarcely a month in which to rest the crew, overhaul, refit and provision the vessel. The next voyage ended in 13 months and yielded $10,407. in sperm and whale-oil. The *Falcon* was docked from September 30th to January 11, 1865 when Captain Holmes again took her out to the Atlantic whaling grounds. She was absent for one year and nine months and imported a cargo worth almost $30,000. Yes, whaling was still profitable for those who risked it.

However, American whaling itself was nearing the period of its final disintegration. Many factors served to defeat an industry which had proved its strength. Just two years before the oil gusher appeared in Pennsylvania as a formidable rival to the whale, New Bedford in 1857 reached its maximum with 329 vessels; this represented twelve million dollars in the value of the time and gave employment to ten thousand seamen. The introduction of petroleum drove whale-oil and sperm prices down. While the whalebone's value was decreased by the advent of the steel spring. War, too, emerged as an enemy of whaling.

Whaling continued with lessened vigor during the 1850's and throughout the Civil War years. The Confederacy was ruthless in its attacks upon and intent to wipe out this Yankee industry. The nation lost at least fifty of its whaling fleet during the war years as a result of privateering by Confederate vessels. The most deadly was the *Shenandoah*. It was responsible for the destruction of 25 New England whalers and the capture of four which they pressed into use as transports. The story of the burning of the Yankee whalers was told by Cornelius E. Hunt, one of the officers of the *Shenandoah*. Americans fought out their last battle of the Civil War in the waters of the Russian Far East. Unaware of General Lee's surrender to General Grant and the end of the war, Lt. James Waddell, in command of the Confederate cruiser *Shenandoah* sailed into the remote Okhotsk Sea and the Bering Strait in the North Pacific Ocean and remained there in May and June of 1865 for the purpose of burning or capturing a group of Yankee whalers. Late in June, after the damage had been done, a trading vessel arrived from San Francisco, having left that port on April 19th. The newspapers it brought proved that the war had terminated two months *before* the Yankee whalers were destroyed.

The Shipping List for 1862 declared that the "Southern pirate Semmes has already made frightful havoc with whaling vessels and his piratical ship, the *Alabama,* threatens to become the scourge of the seas." The *Shenandoah, Florida* and *Alabama* continued to prey upon New England whalers throughout the Civil War.

In addition to participating in frigid Arctic wastes in the ultimate struggle at sea between the North and the South, Yankee whalers played a unique role in assisting the Union cause. An early method of the Federal authorities for the restoration of the Union was the blockading of Confederate ports. This meant that 185 harbors and river openings were to be closed against commercial sailings. Foremost was the blocking of Charleston, the leading port on the Atlantic coast, and Savannah. The federal Navy Department purchased forty-six old whalers, loaded them with stones after they had been prepared for sinking by a series of plugs, and sunk them off the harbor of Charleston, South Carolina, and also off the harbor of Savannah, Georgia, ninety miles southwest of Charleston. The plan was full of irony for the dying whale fishery.

On November 15, 1861 what has come to be known in history as the *Stone Fleet* sailed out of New Bedford Harbor fifteen strong to meet a pre-ordained fate. There were other sailings of this kind and sinkings. One of the Stone Fleet was the Bark *Messenger* of Salem, noted as becoming "one of the Stone Fleet No. 2 sunk in 1861." The *Messenger* had sailed for Benjamin Webb in 1854, 1857 and 1860 and on returning from its last whaling voyage June 17, 1861 was sold to Boston for the federal government's use.

The Civil War cut America's whaling tonnage in half. A great change was due in whaling methods, which spelled the end of the old time whaling traditions. Modern whaling can be placed at 1868 when a harpoon-gun was successfully introduced which revolutionized the industry. For the first time the capture of the deadly finback, a rorqual whalebone whale, was made possible. Since 1850 the steam *catcher* was in process of development. By the combination of steam catcher and harpoon-gun operated from a swivel on deck, the risky chase and capture by crews in open whaleboats was outmoded.

On October 3, 1865, Mr. Osgood's Schooner *Para* sailed under Captain Benjamin R. Hussey on a two-year voyage. Captain Hussey secured no sperm oil, but the whale-oil and humpback oil were worth about $13,000. The master of the *Para* bears the same name as the master of the *James Maury,* which sailed in 1841. Captain Benjamin R. Hussey died at sea on the *Maury,* June 15, 1844. The Captain Benjamin R. Hussey of the *Para* voyage twenty-three years later may be the earlier Captain's son.

On November 26, 1866 the *Falcon* sailed under a new master, Captain Macy, and was absent until April 21, 1868. We do not know the results of the voyage as she returned with only 6 barrels of oil. She either disposed of the cargo in a South American port or sent it home ahead. Mr. Osgood secured the *William H. Shailer,* a bark which had formerly been in the African trade, fitted her out and sent her to the Atlantic whaling grounds under Captain Marshall. After an absence of one year, the vessel returned with 100 barrels of sperm. At this time sperm prices had risen until a gallon brought $2.55, so this small cargo was worth $7,650. On December 26, 1867 we find this whaler sailing again for Mr. Osgood and the customs report indicates that she forwarded 243 barrels of sperm and 20 barrels of whale-oil. The *William H. Shailer* never returned to Salem. Condemned as unseaworthy, her remains were sold at Rio Janeiro on November 6, 1869.

In 1867 the United States maintained a fleet of 170 sailing whalers with 62 from New Bedford and 38 from Provincetown. Four years later, the over-all total had been reduced to 126 whalers, with 20 freighters operating as whalers from New York. In 1869 Nantucket sent out its last whaler the Bark *Oak* which brought to a close the long rivalry between the island port and New Bedford, and with it the best years of whaling itself. Nantucket's long career was characterized by gameness and skill. During the Revolution, Nantucket had to whale or starve. From 1775 to 1783 the island became the only port which attempted to carry on the whaling industry. At the close of the war, 134 of its vessels had fallen into the hands of the English and 15 had been lost at sea. During the War of 1812, Nantucket, thirty miles at sea, was cut off from communication with the mainland by the British Blockade and lost many vessels.

After the war, Nantucket rebuilt its fleet and tried to keep up with New Bedford. By 1820 the size of whalers had increased to the point where the island port had to use floating drydocks or *camels* which ferried the heavier vessels over the shallow bottom of the harbor. By 1850 New Bedford assumed the leadership, which it was never to relinquish. The lengthening of voyages and the enlarging of whalers were the cause of Nantucket's decline rather than the reason sometimes given—that Nantucket ran out of luck when she followed a mistaken policy of "sperm or nothing."

In 1867 there were three whalers registered out of Salem. The *William H. Shailer* was one and the two others were the Brig *Para,* Captain Worth, which had been altered from a schooner to a brig for the voyage, and the Bark *Said Bin Sultan,* Capt. James W. Holmes, which had been built in Newburyport in 1861. The *Para* cruised in the Indian Ocean and gave a fair account of herself. By the time the *Para* returned from a three and a half year voyage in the spring of 1871, the price of Sperm had fallen from $2.55 to $1.31, making her cargo of 876 barrels worth $34,426.80. She had shipped home 116 barrels of sperm and it is possible this reached the market in time to enjoy a good price. The competition of petroleum was making itself felt.

As for the *Said Bin Sultan,* she sailed on June 13, 1867 and went first to the Atlantic whaling grounds and then continued on to the distant Pacific. The voyages grew longer as the whales became more elusive. This whaler was absent for four years. Within that period, she sent home 410 barrels of sperm. When she arrived in Salem on the anniversary of her sailing four years earlier, she hailed 294 barrels of sperm and 149 barrels of whale-oil. The sperm was worth $11,554. and the whale-oil price had declined to 64¢ a gallon. While the *Said Bin Sultan* and *Para* were away, Mr. Osgood sent forth the *Falcon* under Captain Richmond on June 9, 1868. The *Falcon*'s voyage lasted for three years. The importation was 471 barrels of sperm, 4 barrels of whale-oil, and 25 barrels of sperm shipped home.

The great mammal had been hunted with so much zest that his numbers declined and it was no longer possible to find the numerous herds of Sperm and Whalebone whales. Around the 1840's the fishery for the spermaceti whale was conducted off the coast of Chile, Peru and California and in various parts of the Pacific Ocean about Galapagos and Marquesas Islands as well as in the Indian and China Seas, particularly about the Island of Timor. The Sperm whale, travelling in schools, does not frequent either Polar region, but is commonest in tropical and sub-tropical seas. The lush years of the sperm runs seemed to be over.

In 1868, the famous whaling Port of Hull in Britain, sent out its last whaler, the *Truelove,* which had been built in Philadelphia 104 years before and captured by the English during the Revolution. Engaged in whaling since the end of the sixteenth century, the reason given for the decline was the scarcity of the mammals. With Hull out of the running, Dundee, Scotland, became the final whaling port of Britain. Even after the Greenland industry fell on evil days, stout whaling vessels were built at Dundee and as late as the period 1893 to 1911 right whaling was still undertaken by an average of 7 or 8 vessels from there. Finally in 1912 a single whaler sailed from the last of British ports.

The industry as a whole suffered from increased overhead expenses of carrying on the whale fishery which now required prolonged voyages to overtake surviving whales. As we have seen, Salem voyages stretched into three and four years. There was the ever-present problem of securing seamen. Other industries attracted the crews away from whalers as manufacturing developed. More than anything else, the increasing scarcity of whales in the established whaling grounds placed American whaling beyond the pale of profitable enterprise. Whaling did not cease altogether for many years and some American ports continued to whale by sailing ships and by hand harpoon methods into the twentieth century, refusing to believe that the once buoyant institution could die. When the perfection of the harpoon-gun of the Norwegian Svend Foyn had removed men from the peril of personal conflict with the whale in a fragile boat in an open sea, the old sailing ships themselves were heading toward the setting sun. Steam had begun its complete conquest of navigation.

The year 1871 removed many whalers from New England with great losses. The entire Arctic fleet was destroyed when 34 vessels were crushed in the ice and had to be abandoned as a total loss. The men were rescued and over 1,200 seamen were in Honolulu without employment. The loss in vessels, cargoes, wages and equipment is placed at about two million dollars. New Bedford alone lost 21 vessels and sustained a loss of one million dollars. In a similar disaster 12 whalers were lost in 1876 with the loss of fifty lives. In 1888, five vessels in the Arctic were driven ashore by storms and in 1897, eight more whalers were caught and crushed in Arctic ice.

But let us return to the Salem whalers sent out by Mr. Osgood, all of which reported back in 1871 to close Salem's whaling chapter. Both the *Para* and *Said Bin Sultan* left Salem in 1867. The *Falcon* left six months after the *Para* and a year after the *Said Bin Sultan*. All three vessels reported home within a few weeks of each other. As it happened, the *Falcon* was the first to dock—on May 9, 1871. One week later, May 16, the *Para* sailed into Salem Harbor. About a month later, June 13, 1871, the *Said Bin Sultan* returned, and became Salem's *last whaler*. This examination of the record rules out the *Messenger* which is thus classified by Starbuck. His reference to her in these terms must be a misprint as the *Messenger* left Salem on her final whaling cruise in 1860 and was sold to Boston the following year which was *ten* years prior to Salem's final whaling year. The *Messenger* was sunk in 1861 which was *six* years before the *Para, Said Bin Sultan* and *William H. Shailer* left Salem on whaling voyages and seven years before the *Falcon* sailed in 1868.

The last whalers registered in Salem were not allowed to lie idly at its wharves for years on end, the fate of so many whalers retired from service. Both the *Falcon* and *Para* were promptly sold to Boston within the year they returned and the *Said Bin Sultan* followed them there in 1872.

An interesting theme is the fate of the few Salem whalers which survived shipwreck and other disasters. Some were sold to other ports and others were converted into commercial carriers, finishing their sailing careers in a more prosaic pattern than that pursued in the romantic whaling years. A nostalgic tale came to light about the Ship *Bengal,* built in Salem in 1816, and owned by Pickering Dodge. The *Bengal* was outfitted as a Salem whaler and achieved four voyages between 1832 and 1840. On July 24, 1840 the *Bengal* was registered with John B. Osgood as owner and Benjamin Jackson, master. She was sold to New London in 1844 and later became known as the whaler *Chance.*

For many years the original *Bengal* was in use as a New Zealand cargo carrier. In 1886 she was finally withdrawn from Lloyd's listing and her sailing days were over. The old vessel then merited her existence by becoming a hulk used as a floating storehouse at Bluff Harbor, New Zealand. As she had a tendency to leak, windmills were set up on board to run the necessary pumps. When the leaks grew too serious for the pumps to handle, the *Bengal,* whose name by this time was long forgotten, was finally abandoned even as a hulk. A faded snapshot in the files of the Peabody Museum in Salem, sent in by F. G. Layton of Wellington, New Zealand in 1911, shows the old whaler as she appeared after being abandoned and beached. One mast is missing and she is listing badly, but she still retains her characteristics as a whaler. The ultimate fate of the old *Bengal* is unknown but it is not difficult to surmise.

In 1906 New Bedford still had 24 whalers active in the industry. Provincetown registered some whalers and after the lapse of 70 years a single whaler appears on the customs records of Norwich, Connecticut. Whalers continued to sail from New Bedford during the first quarter of this century. At the time of World War I, sperm oil brought $1.35 a gallon. The *Charles*

W. Morgan sailed to the Antarctic region and returned in a year with a cargo worth $53,000. There were other such voyages, but the end was not far away.

During the summer of 1920, the eighty-year old *Charles W. Morgan* was again fitted out for a whaling voyage. Some whalers were prepared for a whaling cruise which they never achieved. There was the square-rigger *Wanderer* which anchored off Mischaum Point, New Bedford, on August 25, 1924, preparatory to an early morning departure. A near-hurricane had other plans for the *Wanderer*. It lashed at her sails and masts with such relentless ferocity, the vessel was wrecked at Cuttyhunk. The summer of the following year saw the New Bedford Schooner *Margaret* return safely from a whaling voyage. In 1925, New Bedford sent out its last whaler.

Perhaps this is as good a time as any to clear up the question of which American vessel was the last to sail in pursuit of whales. Although the *Charles W. Morgan* is the last surviving whaling vessel, the *Morgan* did *not* sail on the last voyage. Some confusion arises from the letter which George A. Grant, Pacific Club, Nantucket Whaling Museum, wrote in May, 1931, to Captain Everett Joshua Edwards, author of *Whale Off!* Capt. Grant held that the last whaler that sailed from New Bedford was the Bark *Bertha* which he alleged was "lost going out on Cuttyhunk Island in 1926."

A letter of inquiry directed to the late William H. Tripp, Curator of the Whaling Museum at New Bedford evoked from him the prompt and doubt-dispelling declaration that "The very last whaling voyage was made by the Schooner *John R. Manta* to the Hatteras Whaling Grounds in 1925, any other statements to the contrary." Mr. Tripp advised further that "Capt. Grant, whom I knew very well, was greatly mistaken in what he wrote to Capt. Edwards. The Bark *Bertha* was sold to the Portuguese for a packet in 1917 and was lost at sea in 1918. What Capt. Grant may have had in mind was the Bark *Wanderer* which sailed from New Bedford on a whaling voyage in August 1924 and two days later was wrecked on the Island of Cuttyhunk. It was eight months later when I sailed on the *John R. Manta,* and there have been no (sailing) whaling voyages out of the United States since that time." The experiences of Mr. Tripp on this final voyage have been graphically described in his book *There Goes Flukes!* published in 1938.

The veteran *Charles W. Morgan* was used in several motion pictures in the 1920's. In 1921, she appeared in the World Film Corporation's production "Miss Petticoats." In 1923, the old whaler was the central point of interest in the film "Down to the Sea in Ships" by the New Bedford Whaling Film Corporation produced by Elmer Clifton. Following this, the *Charles W. Morgan* was fully rigged and taken to Salem, Massachusetts for use in the film "Java Head" which was based on Joseph Hergesheimer's novel about General Frederick Townsend Ward of Salem, produced by Famous Players-Lasky Corporation now Paramount Pictures.

After serving the purpose of the film-makers, the old ship was taken to Fairhaven where it was forgotten. Without attention, the *Morgan* deteriorated into a derelict. Ultimately she was rescued by a group of New Bedford men who were successful in interesting the wealthy Colonel E. H. R. Green (son of Wall Street's Hetty Green). It appears that the Colonel's grandfather had once owned the *Morgan.* At any rate, negotiations over, she was towed to Colonel Green's Round Hill estate on Buzzard's Bay where bereft of her billowing sails, the *Morgan* became a monument to the men and ships of America's great whaling years.

When Colonel Green died in 1936, the *Morgan* lost her main benefactor. Two years later the vessel barely survived a savage lashing by the September, 1938 hurricane. It would have taken a good deal of money to restore the damage done. For a few years, it looked as if the *Charles W. Morgan* would be dismantled and her now-unsightly hulk removed. In a pitiful condition, she was finally turned over to the Marine Museum of Mystic, Connecticut, in 1941, just one hundred years after her maiden voyage of 1841. The *Charles W. Morgan* completed 37 whaling voyages and earned more than $2,000,000 for her owners.

A visitor on this last survivor of the American whaling fleet might reflect as he steps aboard through the starboard gangway that over these boards the blubber and bone of some 2000 whales were dragged slowly and laboriously inch by inch by a succession of more than 1,500 whalemen during the long career of the *Morgan*. On her foredeck are the fireplaces and the great iron try-kettles where the blubber was boiled. Inside the massive wooden hull are the officers' quarters and the captain's cabin. Up forward in the fo'c's'le are the plank bunks polished and worn by generations of sailors.

A few years before New Bedford surrendered, off-shore whaling was still pursued on Long Island, New York, by the East Hampton and Amagansett men whose ancestors were the first in the New World to prosecute whaling by organized efforts. It will be remembered that the eastern end of Long Island was the scene of the first whaling on this continent where in 1640 Southampton on the southern shore was founded as an offshoot colony from Lynn, Massachusetts, once part of Greater Salem. In 1644 the town was divided into four wards in each of which men were appointed to watch for and take care of the whales that drifted ashore. The perilous off-shore whaling which the early English settlers evolved along the Atlantic coastline was continued by lineal descendants with an almost ritualistic ferocity until around 1910.

By 1910, the North Atlantic Right Whale had almost completely disappeared from these waters. For many years its capture had ceased to be profitable. The final Right Whale taken on the Eastern seaboard was killed by Everett Joshua Edwards in the summer of 1918 six miles off the Amagansett Beach, and has long been an exhibit of the American Museum of Natural

History in New York City. Captain Edwards' home is in Easthampton, which includes Amagansett and is part of Sag Harbor, once a major whaling port. Easthampton, incidentally, was the home of James Loper, whom the Nantucket Islanders in 1672 invited over to teach them the art of whaling. When this book was prepared in its first edition in 1950, there was at Easthampton and Amagansett a small group of retired whalemen in their seventies and eighties, many of whom have now gone to the Valhalla of the whaling grounds. A whaleman was toughened and strengthened by a whaling cruise, imbued with a dedication to the task in hand, trained in the crafty use of his skills, and bore within his character for the rest of his days cunning and courage in meeting life's situations which he had learned in the whaleboats.

CHAPTER

22

Salem's Blubber Hollow

"In every rank, both great and small, it is industry that supports us all."

John Gay (1685-1732)

Very little remains of the whaling era in Salem. Some twenty-three whaling logs of Salem whalers have to date found their way to the marine files of the Essex Institute. With some difficulty a few prints were brought to light of forgotten Salem whalers. The Essex Institute has some old Betty lamps and a garden variety of whale-oil lamps in use before the day of the wax candle and petroleum. But the old records tell of whale-oil refineries and candle factories in Salem.

Caleb Smith, who processed fish oils a century ago, operated a sperm oil and candle factory in 1835 at the foot of Harbor Street close to the wharf owned by Pickering Dodge. This was once near Stage Point, where the Naumkeag Mills have been located to this day. In 1836, Col. Francis Peabody, then thirty-five, commenced in South Salem the refining of whale-oil and sperm oil and the manufacture of candles from spermaceti. In a single year, Col. Peabody purchased $100,000 worth of sperm oil and $50,000 worth of whale-oil for this business. His candles enjoyed an enviable reputation both at home and abroad and he is said to have imported the first braiding machines and to have produced the first candles with the braided wick.

A glowing spermaceti candle is a joy to behold. It yields more light than three tallow candles and has four times as big a flame, the clearest and most beautiful from any substance known to nature. Soon their manufacture in New England and their sale amounted to an important industry, while whale-oil helped to feed the vast variety of metal and glass lamps before the era of kerosene and gas light.

Perhaps Col. Peabody is best known for his founding of the Forest River Lead Company and his manufacture of linseed oil begun at Middleton around 1837 where four years earlier he had established a paper mill. For the processing of the linseed oil, he imported to Salem large quantities of flaxseed from Europe and Calcutta, chartering the Ship *General Harrison* in 1841 for this purpose, and purchasing the Ship *Isaac Hicks* and the Ship *New Jersey,* said to be one of the largest ships ever to discharge a cargo at Salem, displacing between 600 and 700 tons. Col. Peabody's enterprises were always on an ambitious scale, but he is also remembered for other talents. At twenty-five, before he became involved in profits from Moby Dick, he had organized a great sham battle in Danvers in 1826. In the educational field, Col. Peabody had introduced public lectures in Salem and toward the end of his life served as a very active president of the Essex Institute from 1865 to 1867.

Another sperm and whale-oil refinery is said to have been early located in the section on Boston Street which has been Salem's tanning and leather center since 1800. There is a legend in Salem that tanners used whale-oil for dressing leather. About a century ago a factory off Boston Street prepared sperm oil for lamps and also made it into Spermaceti candles. A search of the Salem Directory of 1854-55 disclosed that S. W. Robertson & Co. were soap and candle manufacturers at 44 Boston Street. It is felt that this may have been the company referred to. The Directory of 1856-57 indicated that oil and candles were produced by the Salem and South Danvers Oil Company on Mason Street and also by Ebenezer Seccomb at the foot of Harbor Street.

The particular area between Boston Street and Looney's Hill lying right between Salem and Peabody is called *Blubber Hol-*

low. There was a time when one-third of all the leather purchased in North America came from vats within a mile or so of Blubber Hollow. By association this section has come to be synonymous with leather. So obscure is its origin that a thorough search of available sources and a check with Salem antiquarians failed to reveal any clue on just how Blubber Hollow received its name. The question remains open. We do know that *Knocker's Hole* at High Street in Salem was named for the hammering of shipwrights building vessels some years before the Revolution. At that time the tide flowed up into Creek Court and along the bottom of High Street, making it possible to launch the vessels when completed. Is it not possible that whale-oil refineries, coopering and other activities connected with the whaling industry existed in sufficient numbers in Blubber Hollow to give the section its name?

A whale made a brief call at Salem Harbor on July 13, 1948 as a reminder that it was once his custom to frequent these shores. "A frisky male" reported the *Salem Evening News* "from all accounts one of the largest ever seen off this coast, rose and spouted within close view of many spectators." About a year later, June 23, 1949, the *News* again reported that "a large whale was sighted in the waters off Marblehead by Harbor Patrolman William Goodwin as he was fishing about one and a half miles off-shore near the Master Orne fishing grounds."

From time to time a whale finds himself in the waters surrounding Salem, thrashing about with its great tail and spouting to the astonishment of everyone, but these polite calls at Salem are purely symbolic. The whale has no intention to return in the numbers which once merited the consideration of Charles I, the English monarch, whose Royal Charter of 1629 to the Puritans of the Massachusetts Bay Colony included a provision for catching whales. Salem's whaling day is done offshore and at sea. All that remained was the telling of the tale which should long ago have become a part of the American legend.

CHAPTER

23

Modern Antarctic Whaling

Ho! Ho! Ho! We're whalemen all
And a terrible life we lead:
We hunt in the ice-packs gloomy pall
That suburban folk may feed!
We hunt and we bomb
And the whale spouts blood
And never our homes do we see,
But the peak of our manhood's at the flood,
Ho! Happy mariners we!!
Alan John Villier's *Whalers of the Midnight Sun.*

While the whale hunt of the square-rigged sailing ship is past, whales are still fair game by modern methods. As early as 1830 a harpoon gun of sorts was used, but not until 1868 was it perfected. The introduction of the harpoon gun fashioned by the Norwegian Svend Foyn was the turning point in revolutionizing the whaling industry. It occurred as the era of sailing whalers and open whaleboats drew to a close superseded by steamship and mechanical improvements in the pursuit and catching of whales.

Norway claims as the father of Norwegian whaling Ochter who came from Helgoland and whose feats made him the first recorded whaleman. Another Norwegian, Svend Foyn produced the means of modernizing the ancient industry. And still another Norwegian, Captain C. A. Larsen, was to initiate the important Antarctic whaling enterprise on an impressive scale.

While Sperm and Right whales grew increasingly scarce in the North and South Atlantic, a small American fleet still continued to hunt in these waters before the turn of the century. After 1900, however, the Antarctic became the most productive whaling theater. Fresh impetus was provided when new whaling grounds were discovered in the sub-Antarctic seas in 1904 with thousands upon thousands of whales spouting and frolicking among the icebergs.

As attention turned to the still virginal waters of the Antarctic, Captain C. A. Larsen aroused the financial interest of the Argentines. They extended funds which helped him to found a company in 1904, which was the first to develop the industry to any extent in the Antarctic. Captain Larsen was responsible for triggering the greatest and wealthiest whale fishery in the long history of whaling. His locale of activity was the Falklands and the bleak South Georgia Islands in the South Atlantic. It is interesting to recall that a half century earlier, Salem sealers and whalers were active here. In the desolation of ice and snow, Captain Larsen and his men found their abandoned equipment and try-pots, in such surprisingly good condition, the Norwegians promptly pressed the old gear into service.

In South Georgia, the whole technique of modern whaling was evolved. In 1908 another company, organized by Norwegians, began to whale from a base at Durban, South Africa and in the next three years a number of companies operated from shore stations on the African coast.

Up to the present, there have been three distinct periods in the whaling industry.

1. From the time the Basques began to whale in the twelfth century until 1868, the hand harpoon was used in personal combat with the whale.
2. The development of the Norwegian harpoon gun in 1868 speeded the kill and increased efficiency in hauling whales to shore stations for processing.
3. Modern or pelagic whaling began with the development of

factory ships which processed whales at sea and employed steam catcher boats.

There have been three distinct types of vessels employed:

1. Boat or off-shore coastal whaling.
2. Sailing vessels or American whaling.
3. Steam or Norwegian whaling.

By the time the Norwegians were soundly established in the Antarctic, New England saw its surviving whalers check into port for the last time; the prolonged domination by Americans of what had been the most picturesque of domestic industries was transferred to another nation. By 1928 the Yankee had conceded supremacy to the Viking. The center of the industry shifted from the North Atlantic to the Falkland Islands, the South Shetlands, South Orkneys and South Sandwich Isles. In 1929, the Norwegians revived the industry by means of stock companies and three of the larger companies reported combined profits of $2,000,000 with 30% dividends paid on stock.

The phenomenal development of the Antarctic industry in the twentieth century recalled the Greenland whaling of the early seventeenth century when as many as fifty sailing vessels worked within the sight of each other near the North Pole. As many as a thousand vessels a year visited the Arctic region. While the number of ships to the Antarctic was impressive, it never matched this number. In this activity, the Norwegians were definitely in the lead. The 1928-29 South Polar season saw a total of 25 shore stations, 30 great floating factory ships and 237 killer boats in action. Norway's catch was 14,996 whales; Britain secured 8,230; Denmark and Argentina together secured 1,770 whales.

In 1930 the shore whaling stations in South Africa and in the Antarctic were operated by both Norwegian and British companies. Since the onset of pelagic or factory whaling, many nations were represented in the Antarctic. Holland, Germany and Japan entered ships, followed by Russia on a small scale, with New Zealand, South Africa and Australia taking an increasing interest. Once a leader in the whale hunt, the United States did not figure in the season of 1930; the limited amount of whaling off our West Coast and at Alaskan shore stations were under Norwegian auspices, as was also the station at British Columbia.

In 1931 there were 31 floating factory ships of all active nations with a complement of 210 whale catchers. The greatest number were owned by Norway who was active with 11 factory ships and 77 catchers. Japan was active with 4 factory ships. Of the 8,399 men working in the Antarctic 7,145 were Norwegian, 763 were German, 362 were British and 129 of other nations.

Up to 1908, the industry of the sailing ship took more than 50% of the declining number of whales in the Northern hemisphere. Following the whales to their habitat, by 1938 over 90% were taken from the Northern hemisphere. A thirty year span served to relocate the whaling theater. In 1935 Norway controlled two-thirds of the world's whaling. In the 1938-39 season, Britain and Norway were neck and neck in the race for top position in the lucrative whaling industry. The world whaling fleet comprised 39 factory ships, 14 land stations and 353 catcher boats operating with a personnel of 12,000 men. Two of these factory ships flew the American flag.

World War II curtailed whaling operations in the Antarctic by withdrawing the chief operators. During the war, governments learned to stockpile the oil as edible fat and it promptly became a top priority item.

As a post-war Europe looked to the whale to supply the greater part of its fat needs, the 1945-46 hunting season saw only a remnant of the large pre-war fleet report for operations, three British factory ships and six Norwegian. England's critically small supply of edible fat was badly in need of bolstering. The half a million tons of whale-oil the country had hoarded against U boat action was gone by the time the whalers returned to Antarctic activity.

Post-war whaling by Britain was a government matter. The three factory ships operated under the Ministry of Food and a government Food Board allocated the oil secured. Europe anticipated receiving 135,000 tons of needed whale-oil. The whalers failed by one-fourth to achieve this goal. Britain's Food Ministry hoped for a harvest of 75,000 tons of whale-oil from the first post-war cruise. The three factory ships returned with a little over half, or 40,000 tons, with the bad news that whales were not as plentiful as hoped. The failure to meet the quota set caused a cut in fat rations in England.

In the 30's the ceaseless whale hunt threatened to exterminate the whale in every ocean. During the inter-war period, the demand for whale-oil increased and gained a new importance as the list of the uses of whale products lengthened because of technical advances. Before the war, the value of whale-oil fluctuated between $40 and $76 a ton. In 1935, whale-oil had reached over $400 a ton. In the 1945-46 season, a ton of whale-oil was worth over $180. *Science Digest* placed the value at $250 a ton. Each barrel weighed 400 pounds, making 5 barrels to the ton.

After the war, four nations were vitally interested in the whale fishery, Britain, Norway, Germany and Japan. In the 1946-47 season, 15 factory ships were in the Antarctic: 3 from the United Kingdom, 7 from Norway, 1 from Holland, 1 from South Africa, 2 from Japan and 1 from Russia, with Germany not yet resuming activities. The number of whales taken by this world fleet was 30,000.

In the 1947-48 season, Norway entered two more floating factories increasing the total to seventeen factories operating in the Antarctic. The number of whales taken rose from 30,000 to 31,000. As a Blue Whale yields some 50 tons of oil, production was at a very high level. According to the Office of Foreign

Activities, United States Department of Interior, there were in 1950 about 25 shore stations and 19 factory ships operating in the world, with a take of around 30,000 whales annually.

Half of the whale-oil secured in the Antarctic found a market in the United States, the other half going to Europe, yet the stake of the United States in the world industry was negligible. American interests were at a low ebb, due to high labor costs, competition with cheap vegetable oils, substitutes for sperm oil in the lubrication field and no pressing need for the meat for human consumption.

Since the days of the Puritan fathers, Americans have taken a greater or lesser part in the whaling industry, never completely relinquishing its interest. In the 1930's American interests operated a shore station at Akutan, East Aleutians; Port Hobron, Gulf of Alaska; Port Armstrong, Southwest Alaska; and Naden and Rose harbors in the Queen Charlottes. There was also a shore station at Trinidad Head, just north of Eureka, California. From 1935 to 1938 American interests in New York chartered 1 to 2 Norwegian factory ships for operations against Humpbacks in the Antarctic and Australia.

In 1939 a small United States shore station was built in northwestern California at Fields Landing, six miles south of Eureka on Humboldt Bay. It began operation in 1940. Meanwhile in the following year, the station at Akutan and the Queen Charlottes closed down and the factory ship employed was dismantled. Fields Landing, however, met with modest success under the operation of Maritime Industries which reported these catches:

1940	29 whales	1945	not operated
1941	24 "	1946	" "
1942	26 "	1947	38 whales
1943	29 "	1948	67 "
1944	5 "	1949	no report ceased operation

Inquiry in December of 1961 of the United States Department of the Interior, (Fish and Wildlife Service, Bureau of Commercial Fisheries) Washington, D.C. elicited response of

Mr. John I. Hodges, Chief, Branch of Resource Management, to the effect that the whaling station at Fields Landing, Eureka, California was closed down "some years ago." However during the 1961 season, three land stations operated in the United States, two at Richmond, California and one at Warrenton, Oregon. Six American catcher vessels were utilized in taking 316 whales.

Whaling has its seasons fixed by the Commission governing world whaling with headquarters in Norway. Operations for the various national groups of expeditions for last season (1959-60) were as follows:

		Catching days
Norwegian	December 28—April 7, 1960	102
British	December 28—April 7 "	102
Japanese	December 28—March 26 "	90
Soviet Russian		
Sovietskaya Ukraina	December 28—March 31 "	95
Slava	December 28—March 27 "	91
The Netherlands	December 16—April 15 "	122
	Average number of days 99	

During the Antarctic season 1959-60 there were five countries which carried on pelagic whaling with 20 floating factories and 220 catchers. This whaling matérial and that which was used in the preceding season belonged to the participating countries:

	Season 1959-60		Season 1958-59	
	Floating Factories	Catchers	Floating Factories	Catchers
Norway	8	70	9	93
United Kingdom	3	31	3	37
Japan	6	69	6	69
The Netherlands	1	13	1	12
U.S.S.R.	2	37	1	24
Total:	20	220	20	235

Is there a whaling operation outside of the Antarctic? The answer is "yes". In 1959 the following companies were involved in such operations:

Société des Pécheries Coloniales à la Baleine commenced whaling from a shore station at Cape Lopez, Gabon (earlier called French Congo).

Barrier Whaling Co. Ltd commenced whaling from a shore station on Great Barrier Island, New Zealand.

Company which ceased operation in this season:

Den kgl. gronlandske Handel (The Royal Greenland Trade Department) on the west coast of Greenland.

(Mention should also be made that three of the Antarctic floating factories in 1959 caught 16 sperm-whales in the South Atlantic Ocean on their way to the Antarctic.)

In 1959, whaling was prosecuted from the following grounds outside of the Antarctic:

Africa. Natal, Cape Province, Gabon (earlier French Congo)

Brazil. Paraiba

North Atlantic and Arctic. Azores, Madeira, Spain, Norway, Iceland, Newfoundland.

Pacific North. California, British Columbia, North Pacific Ocean and Bering Sea, Kamtchatka, Kuril Islands, Japan, Ryukyu Islands

Other grounds. South Atlantic Ocean, Peru, Chile, Australia, New Zealand.

Men in the Antarctic

During the three latest seasons, a survey was made of the nationality of the crews employed on the freezing and refrigeratorships operating in the Antarctic with these results:

Number of men employed in the Antarctic whaling industry

	Season 1957/58		Season 1958/59		Season 1959/60	
Nationalities	Number	Per cent	Number	Per cent	Number	Per Cent
Norwegian	6,693	37.6	6,817	38.9	6,152	34.5
Japanese	8,402	47.2	7,974	45.4	8,100	45.4
British	911	5.1	944	5.4	890	5.0
Russian[1]	1,020	5.8	1,020	5.8	1,900	10.7
Other[2]	765	4.3	792	4.5	787	4.4
Total	17,791	100.0	17,547	100.0	17,829	100.0

1) Estimated 2) includes also men whose nationality is unknown

The total catch of whales has increased in recent years to incredible proportions. In the season of 1957-58 about 64,500 whales were taken and in 1958-59 the figures also reached the largest ever recorded. The following table supplied by International Whaling Statistics (XLVI Oslo 1961) edited by The Committee for Whaling Statistics concerns essential data for the Antarctic.

Whaling on grounds outside the Antarctic during the year 1959 was carried on from 53 shore stations, 3 floating factories and with 164 catchers. This is the same number of floating factories, but one shore station less and 4 catchers more than in 1958. The number of shore stations operated on grounds outside the Antarctic in 1958 and 1959 were the highest ever recorded, and only in one previous season a higher number of catchers has been engaged outside the Antarctic, namely in 1951 with the maximum figure of 208 catchers.

In the survey below is recorded the number of whales taken on grounds *outside* the Antarctic during the period 1937-38 to 1958-59 with figures for the Antarctic whaling included in order to give the total world whaling results during this period.

The total catch on the grounds outside the Antarctic has increased considerably during the postwar seasons. The increase

Season	Total of whales killed	In Antarctic			On other grounds		
		Total	Baleen whales	Sperm-whales	Total	Baleen whales	Sperm-whales
1937-38	54,902	46,039	45,172	867	8,863	5,967	2,896
1938-39	45,783	38,356	35,771	2,585	7,427	4,501	2,926
1939-40	37,709	32,900	30,962	1,938	4,809	2,076	2,733
1940-41	23,638	16,363	15,559	804	7,275	2,438	4,837
1941-42	8,014	1,425	1,316	109	6,589	1,741	4,848
1942-43	8,372	998	974	24	7,374	1,895	5,479
1943-44	6,445	1,799	1,698	101	4,646	2,133	2,513
1944-45	5,919	2,891	2,846	45	3,028	1,404	1,624
1945-46	19,384	13,387	13,114	273	5,997	2,809	3,188
1946-47	34,820	25,593	24,162	1,431	9,227	3,112	6,115
1947-48	43,382	31,318	28,696	2,622	12,064	4,836	7,228
1948-49	44,002	31,435	27,357	4,078	12,567	7,629	4,938
1949-50	45,093	32,396	29,669	2,727	12,697	7,205	5,492
1950-51	55,812	33,997	29,029	4,968	21,815	8,502	13,313
1951-52	49,816	35,237	29,752	5,485	14,579	8,506	6,073
1952-53	44,988	30,653	28,321	2,332	14,335	7,090	7,245
1953-54	53,642	34,872	31,993	2,879	18,770	8,106	10,664
1954-55	55,062	37,654	31,864	5,790	17,408	7,605	9,803
1955-56	58,126	38,580	31,606	6,974	19,546	7,930	11,616
1956-57	59,056	36,115	31,686	4,429	22,941	8,214	14,727
1957-58	64,586	39,896	33,361	6,535	24,690	9,379	15,311
1958-59	64,489	38,890	33,238	5,652	25,599	9,953	15,646

Total Antarctic

Years	Species of whales caught						Total of whales	Oil production	Expeditions			Number o whales per boat
	Blue	Fin	Humpback	Sei	Sperm	Others			Shore stations	Floating factories	Catchers	
								Barrels				
1919-20	1,874	3,213	261	71	8	1) 14	5,441	272,817	6	6	44	124
1920-21	2,617	5,491	260	36	31	1) 13	8,448	390,627	6	8	47	180
1921-22	4,416	2,492	9	103	3	—	7,023	452,517	6	8	46	153
1922-23	5,683	3,677	517	10	23	—	9,910	614,547	6	13	60	165
1923-24	3,732	3,035	233	193	66	1) 12	7,271	464,678	6	14	66	110
1924-25	5,703	4,366	359	1	59	—	10,488	697,091	6	13	65	161
1925-26	4,697	8,916	364	195	37	1) 10	14,219	783,307	6	15	70	203
1926-27	6,545	5,102	189	778	39	1) 12	12,665	872,362	6	17	80	158
1927-28	8,334	4,459	23	883	72	1) 4	13,775	1,037,392	6	18	84	164
1928-29	12,734	6,689	48	808	62	—	20,341	1,631,340	6	26	111	183
1929-30	17,898	11,614	853	216	73	1) 1	30,655	2,546,759	6	38	194	158
1930-31	29,410	10,017	576	145	51	1) 2	40,201	3,608,348	6	41	232	173
1931-32	6,488	2,871	184	16	13	—	9,572	808,560	2	5	45	213
1932-33	18,891	5,168	159	2	107	—	24,327	2,456,462	1	17	118	206
1933-34	17,349	7,200	872	—	666	—	26,087	2,395,544	2	19	126	207
1934-35	16,500	12,500	1,965	266	577	—	31,808	2,453,999	2	23	153	208
1935-36	17,731	9,697	3,162	2	399	—	30,991	2,436,338	2	24	175	177
1936-37	14,304	14,381	4,477	490	926	1) 1	34,579	2,658,108	2	30	196	176
1937-38	14,923	28,009	2,079	161	867	—	46,039	3,340,330	2	31	256	180
1938-39	14,081	20,784	883	22	2,585	1) 1	38,356	2,820,771	2	34	281	136
1939-40	11,480	18,694	2	81	1,938	2) 705	32,900	2,544,253	2	28	240	137
1940-41	4,943	7,831	2,675	110	804	—	16,363	1,100,008	1	11	93	176
1941-42	59	1,189	16	52	109	—	1,425	77,819	2	—	12	119
1942-43	125	776	—	73	24	—	998	50,960	1	—	6	166
1943-44	339	1,158	4	197	101	—	1,799	132,001	1	1	15	120
1944-45	1,042	1,666	60	78	45	—	2,891	223,540	1	1	15	193
1945-46	3,606	9,185	238	85	273	—	13,387	818,652	3	9	93	144
1946-47	9,192	14,547	29	393	1,431	1) 1	25,593	1,939,742	3	15	147	174
1947-48	6,908	21,141	26	621	2,622	—	31,318	2,104,051	3	17	183	171
1948-49	7,625	19,123	31	578	4,078	—	31,435	2,219,621	3	18	212	148
1949-50	6,182	20,060	2,143	1,284	2,727	—	32,396	2,166,489	3	18	237	137
1950-51	7,048	19,456	1,638	886	4,968	3) 1	33,997	2,303,690	3	19	260	131
1951-52	5,130	22,527	1,556	530	5,485	4) 9	35,237	2,479,180	3	20	289	122
1952-53	3,870	22,867	963	621	2,332	—	30,653	2,118,097	3	16	251	122
1953-54	2,697	27,659	605	1,029	2,879	4) 3	34,872	2,285,720	3	17	227	154
1954-55	2,176	28,624	495	569	5,790	—	37,654	2,242,555	3	19	254	148
1955-56	1,614	27,958	1,432	560	6,974	4) 42	38,580	2,307,171	3	19	278	139
1956-57	1,512	27,757	679	1,692	4,429	4) 46	36,115	2,246,922	3	20	246	147
1957-58	1,690	27,473	396	3,309	6,535	4) 493	39,896	2,317,638	2	20	257	155
1958-59	1,192	27,128	2,394	2,421	5,652	4) 103	38,890	2,152,659	3	20	256	152
1959-60	1,239	27,575	1,338	4,309	4,227	4) 204	38,892	2,148,438	3	20	241	161

[illegible] 3) Bottlenose. 4) Minke-whales.

in the later years is, however, principally a result of the large catch of sperm-whales. Since the 1949 season the catch of sperm-whales outside the Antarctic has increased from about 5,000 animals in 1949 to 15,600 in 1959. The number of baleen whales killed has since 1949 varied between 7,100 and 9,900 only.

In order to be able to analyse the statistics, the figures must be studied in respect to the various grounds as well as in respect to the species of whales killed.

In the *North Atlantic and Arctic* whaling has been carried on continually during the post-war seasons from Norway, the Faroe Islands, Iceland (from 1948), and Greenland. From Faroe Islands and Greenland, however, there was no whaling in 1959. Whaling from Newfoundland and Labrador was of considerable importance during the years 1946-1951. Later the stations were closed, but whaling has been resumed on a considerably reduced scale during the last years.

The whaling operations off the coast of Norway have been reduced in recent years. Up till 1955 4 shore stations with 11-12 catchers were active, but in 1957 and later whaling was carried on from 2 shore stations with 6 catchers only. The aggregate number of baleen whales killed as well as the number of whales killed per catcher thus declined in 1958 to 129 and 21 whales, respectively. In 1959, however, the corresponding figures increased to 184 whales and 31 whales, respectively.

The baleen whale catch off the Faroe Islands has declined since 1951. In later years the catch has varied considerably from one year to another. In 1956 49 baleen whales were killed, in 1957 142 and in 1958 17.

The catch of baleen whales off Iceland has not shown any decline in the number of baleen whales taken. This does not, however, apply to blue-whales. From and including the summer 1955 the taking of blue-whales in the North Atlantic and Arctic was prohibited for a period of 5 years, but Denmark and Iceland have protested against this prohibition, and are thus free to kill blue-whales. The number of blue-whales killed during the years before 1955 and after that period off the Faroe Islands, Iceland and Greenland indicate that the stock is declining.

From Greenland and Newfoundland whaling in later years has been carried on with one single catcher, and it is impossible to draw any definite conclusion as to the decline of the catch or to the diminishing of the stock.

In later years, whaling has also been carried on in the south eastern part of the Atlantic from the *Azores, Madeira* and *Spain*. This whaling is chiefly restricted to the catch of sperm-whales. The figures recorded for the baleen whale catch have declined.

The whaling off the *African* coast during the latest seasons has been carried on off Natal and Cape Province only, Natal being the only ground continually operated in all postwar years. The catch of sperm-whales has been of considerable importance in these years. As to baleen whales, the catch of humpbacks was of major importance to the whaling off the coasts of Madagascar and French Congo. The activities off the coast of Madagascar stopped in 1950. During that year the catch of humpbacks was 711 against 1,333 in 1949. The whaling off the coast of French Congo stopped in 1952, when the number of humpbacks killed had decreased to 265, from 1,105 in 1951. On the Congo ground the catch of sei-whales was of some importance. In 1959 whaling was resumed from this ground now named "Gabon", and the number of humpbacks taken was 160. The baleen whale catch off Natal in later years has been based upon the catch of fin whales. Next to this species comes the sei-whale. The catch of blue-whales and humpbacks has declined considerably; whaling off Cape Province was resumed in 1957. The results recorded show that most of the baleen whales taken are sei-whales, with the fin-whales next in importance.

In recent years whaling in the Pacific North has been prosecuted from California, British Columbia and the Ryukyu Islands. However, whaling from Bonin Island ceased in 1952.

The whaling station on the coast of British Columbia has been operating in all postwar years. Most of the baleen whales killed on these grounds during the last years have been fin-whales. In some of the earlier seasons the humpbacks and sei-whales killed exceeded the number of fin-whales, but in later years the catch of the former species has declined. In 1959, however, the catch of sei-whales and blue-whales increased.

Sperm whales have been taken from the shore stations on the coasts of Japan. Whaling in the North Pacific Ocean and Bering Sea has been carried on since 1955 from 2 floating factories with 13 and 17 catchers. Whaling from Kamchatka is based on the catch of Sperm whales. In 1957 and 1958 one floating factory with 7 catchers and in 1959 one floating factory with 9 catchers were in operation.

Whaling off the Kuril Islands has been carried on during the last four seasons from four shore stations and with 14 to 22 catchers with most whales taken being the Sperms. Brazil maintains from 1 to 2 shore stations with 2 to 3 catchers and Peru participates with from 2 to 3 shore stations, taking almost exclusively Sperms in the last three seasons. In Chile whaling has been carried on in 1958 and 1959 from three shore stations with 13 and 14 catchers respectively. The whaling off Australia and New Zealand is mainly restricted to humpbacks with some Sperms being taken.

The chief by-products are whale meat and whale meal, of which the total production in 1958-59 were 127,019 tons and 77,795 tons, respectively. Of this quantity 78,658 tons whale meat and 41,284 tons whale meal were produced in the Antarctic.

The whaling industry has developed into a business of highly organized efficiency which has little resemblance to its predecessor. Not even the whales pursued are the same. Old-time whalemen hunted the Right or Baleen whale and the black skinned Sperm, which averaged 50 feet and was a comparatively slow swimmer, doing not more than 10 knots even in a spurt. Men in whaleboats could somehow manage to capture and kill these two species. Long ago Right whales were practically removed and only old bulls of the Sperm family have been found in greatly reduced numbers in the Antarctic. In 1938 less than 5% of the Antarctic catch were Sperms.

Today's whaling is a Rorqual industry. The Rorquals chiefly pursued are Sibbald's Rorqual or Blue whale and the somewhat smaller Common Rorqual or Finner whale. One of the Rorquals hit a speed of 30 miles an hour. This speed was recorded also for a Sei whale. A Humpback is reported to have traveled at 25 miles an hour for a short distance. The giant Blue whales are actually slate blue in color and have sulphur colored bottoms. The Finner and Sei whales are a light gray with glistening white bottoms. Both are fluted or corrugated with deep parallel grooves from throat to navel. Reaching maturity at 75 feet, the male of the Blue whale grows up to 90 feet in length and the female to an even greater footage. Females have been registered over 100 feet. The Finners mature at about 65 feet and the male is larger than the female. Both of these whales of the modern industry take about 12 years to reach physical maturity and then enjoy parenthood for at least 20 years. Their natural life span, adds Sanderson, is not known. Harry B. Bradford says of a Sperm whale that "He may live only 25 years or so." In 1942, the oldest known Blue whale was only 12 years old.

Former whalemen confined their efforts to the Sperm and Right whales which float after they are dead and could be towed back to the mother ship. They left the swift swimming Blue and Finner whales alone. In the first place, it was impossible for a few men in an open whaleboat to capture a mammal of his length and dispatch him with a hand-hurled lance. The Blue whale sinks after he is dead, and earlier whalemen had no apparatus capable of raising a 100 ton weight from a depth of a mile or so after the whale had sounded. Today's harpoon gun injects compressed air into the whale so he won't sink.

Modern whaling first operated from shore stations because

of the need of an abundance of fresh water in processing. Factory ships which could render oil at sea proved more profitable and boilers and condensers were introduced for the production of fresh water from the sea. Whaling supplied a market for used ocean liners as ships as large as 20,000 tons were needed. Around 1930 the British shipyards began the building of whale-oil floating factories. At this time the Norwegians dominated the industry, considerably helped by English capital. The floating factory copied the old pattern of mother ship and several whaleboat chasers. The old whalers tried-out their blubber at sea and remained on cruises for years. Today's pelagic factories can be kept at sea for a year or more and are parents to half a dozen whale catchers which have replaced the old-time rowboats.

Modern catchers look like tugs, weigh up to 350 tons and are about 130 feet long with a speed of 14 knots. They carry a crew of eleven men besides the harpoon gunner who is also master of the vessel. The small catchers overtake a whale. The gunner dispatches him. A pole flying the ship's flag is planted into his shining back to stake out the Company's claim. The chaser proceeds to the next whale and perhaps to the next, if his luck holds out. Later on, the prizes are harvested and towed back to the floating factory. Here, there is enacted a new version of the biblical legend. This time a ship swallows a whale. The big body is drawn up a ramp through a very large opening in the ship's stern. Through this ramp a factory ship can take on board 12 whales a day. By a modern miracle it can convert all 12 whales into oil and other products within 24 hours. Some ships are equipped to take care of twice this number. The figure given in *Scientific Monthly* of June 1938 of "40 whales" a day, however, is doubtful and misleading. Eight years later, or 1946, the maximum reached was one whale an hour, or 24 whales around the clock. This is a far cry from the old days when all hands worked in two shifts around the clock for three days to process a single whale.

A factory ship requires a crew of 180 men and 20 pressure boilers to devour 12 whales a day secured by the whaling fleet which operates with the dead earnestness of a naval squadron. High speed mincing machines with whirling blades can shred so fast that in less than an hour they can slice up the blubber of a whole whale. Giant pressure boilers that use 40 tons of fresh water a day then convert the blubber into oil. To get at every vestige of oil, even the head and the bones are pressure cooked. Thus, once the raw material is secured, science sees to it that it is transformed into commercial products with diabolical alacrity.

Every conceivable device is pressed into the whale war. Whales have been hunted by speedboats, and blimps. Humpback whales have been caught in the channel of Cape Brett, New Zealand in nets 600 feet wide and 200 feet deep. The factory ship *Empire Victory,* 20,000 ton British whaler, carried radar, the all-seeing eye, and echo-ranging gear to the Antarctic to help in tracking down any wary whales and in avoiding dangerous icebergs. Neither the whales nor the icebergs proved cooperative. Results were disappointing. Nevertheless in this 1945 cruise, the first after the war, the *Empire Victory* secured 1,600 whales; 15,000 tons of oil were tried-out aboard ship. Figuring 400 pounds to a barrel, the take was 75,000 barrels, or another measurement sometimes used of 15 barrels to a ton, 225,000 barrels. At the 1945-46 value of $180 a ton, the *Empire Victory*'s cargo was worth $2,700,000.

An outstanding example of a massive whaling ship is the *Balaena.* Truly a behemoth, or biblical mastodon of the sea, the *Balaena,* whose tonnage is given as 21,000 tons, was built in Belfast since World War II for the United Whalers Ltd, London, at a cost of about $6,000,000. The catcher boats are coordinated with the mother ship by radio-telephone. The *flensing* or cutting-in deck of the *Balaena* is slightly longer than an American football field and half as wide. Here 4 giant Blues can be carved up at one time and their blubber removed. The

Balaena has hangars and catapults for three amphibian airplanes which go out to circle the seas to spot whales. They also give weather reports and icing conditions. After power winches have hauled up the whales through the ramp, power derricks and great saws are used in the cutting-in.

The completely electrified *Balaena* carries a crew of 444 men, and 10 catcher vessels. There are a liver-extraction plant, a meat-meal plant, oil purifier plant, deep freeze lockers for There are chemical laboratories, a blacksmith and carpenter shop, a hospital, bakery, laundry, offices and elevators.

The United States participated in the 1950-51 Antarctic season by entering a factory ship of its own. 41 feet were added to a T-2 tanker, giving her a length of 565 feet overall and about 18,641 deadweight tons. The former oil carrier was once manned by 45 men. In her role as a pelagic whale factory, she accommodated a crew of 325 members. The dozen killer boats were former U. S. Navy corvettes which protected convoys during the war. The project was undertaken by the Olympic Whaling Company, Inc. of New York, whose investment on the purchase and conversion of the vessels is in the neighborhood of $8 millions.

Today's chaser or catcher boat is bigger, stronger and faster than before the war. The total whaling fleet of all nations, in which Norwegian and British whalers were foremost, was smaller than before the war. It has become traditional in the modern industry for Norwegians to man the killer boats or chasers, regardless of what nation owns the factory ship. The American whaler *Ulysses* used 8 Norwegian chasers with which they connected on the whaling grounds in the Antarctic. To have built that many chasers in this country would have cost the American company about $3,200,000.

In 1936 the factory crews worked 24 hours around the clock when the catchers were doing well in bringing in the whales. At shore stations in the Antarctic, barracks were squalid and the work went on at an exhausting pace. Some of the men knew about these conditions from experience but were willing to undertake it for a short period. There are always eager farm hands or laborers ready to sign up in Norwegian seaports for a whaling stretch in the Antarctic. It means a chance to earn enough to get married or to buy a farm. Those who take part in the chase, like the whalemen of old, now share according to the catch. They receive a small base wage and a bonus for each whale secured. After the war, working conditions improved but many felt a need of reform throughout the industry. The custom of catching whales and processing them seven days a week is too wearing on human machinery. The season is limited by international agreement. At the base on South Georgia, an island 100 miles long and 15 miles wide, there are now some warm barracks, a small church and moving picture facilities.

Of the personnel, the gunners profit most. Their skill in handling the chaser boats and in aiming the harpoon gun have brought them as high as $20,000 for three months work. Some gunners earn greater sums. There was, for instance, Lars Andersen whose reputation for reckless daring any ancient Viking would have envied. It won for him the accolade of "Lars the Devil." In each three-month season, Andersen managed to earn $50,000. For the balance of the year he lived like a king. War removed whaling from the Norwegian scene and brought invading Nazis. Retiring from whaling just before the war, Lars Andersen died in Trondheim during the German occupation.

The romance and adventure have been tried out of the whaling game. A different kind of skill has replaced the old-time athletic feats of personal combat with the monsters of the deep. This element of sportsmanship has been sacrificed to the spirit of profits. Overtaken by steam catcher boats and killed by explosives, the whale has been reduced to just so much raw material. His huge carcass is now painlessly lifted by 1,000 horsepower winches, his blubber is shredded by electricity and is fed into batteries of enormous pressure boilers on unbelievably large factory ships. What remains is a scientifically planned, precise, methodical enterprise.

Today only the Eskimo pursues his Belugas, or white whales, with the hand harpoons his ancestors used before him and writers in Homeric vein recall the epic grandeur of the Golden Age of whaling. In the general paean, Ivan Sanderson inserts a note of realism. Why lament the not-to-be-recalled Golden Age of whaling? Actually, there have been half a dozen Golden Ages, each connected with a particular kind of whale, and each enjoyed by a different nation. The so-called Golden Age of American whaling, he avers, was not at all golden for those who lived in it.

An aggressive world whaling industry is with us to stay and whales, in numbers beyond the imagination, are being taken in our own times. Few people are aware of its prosecution or importance, while in the heyday of the old industry in America, when some 700 square-riggers sailed out of New England ports, everyone knew of whaling and what it entailed. In point of profits and numbers of whales secured, our own era is the Golden Age for in all of its one thousand year history, the most lucrative cycle of whaling is right now.

CHAPTER

24

Can the Whale Survive?

Nature is the most thrifty thing in the world; she never wastes anything; she undergoes change, but there's no annihilation—the essence remains.

Thomas Binney (1780-1875)

The Blue whale, today's big game, is found near the pack ice of both northern and southern polar hemispheres. Uncontrolled commercial exploitation of the ocean's richest gift almost destroyed the stock of these whales in the northern hemisphere. The 1931 hunting season resulted in a killing orgy which broke all records. The number of whales taken had increased from season to season until in 1931 43,129 whales were killed. This almost equalled the total take of over a century of activity between 1669 and 1778 when the number of whales taken has been estimated at 57,590. When America's whaling industry flourished from 1835 to 1872, some 300,000 whales in all were taken. This means an average of some 8,100 whales a year for 37 years. For the 35 year period from 1904 to 1939, 500,000 whales were killed, or an average of 14,285 a year. Last season over 60,000 were taken.

Modern methods have made possible the taking of vast numbers of the sea mammal beyond all comparison to the days when whaling was a household word. Ivan Sanderson calls attention to the fact that less than 1,000 whales were killed in a 100-year period of whaling by the twenty Basque villages most famous for this pursuit. Nothing in the records of the past can compare to the wholesale destruction of our own times. By 1930 the unrelenting war against the whale was prosecuted by 43 floating factories commonly operating in the Ross Sea area of the Antarctic.

These whaling ships paid tribute to no nation and recognized no restrictions. The British had no restrictions beyond the three mile limit while no power enjoyed a monopoly of the open seas. In the feverish rush for whales in 1931, Britain had her base of operations at South Georgia in the Antarctic, but Norway possessed no Antarctic base. Finally, after a rather bitter dispute with Britain and France, who both claimed the area which Norway coveted, the Norwegians succeeded in securing for a whaling base glacier-covered Bouvet Island, a steep volcanic cone some 900 miles from the Antarctic circle, and a 1000 miles from anything else.

With the whaling industry at a peak, a respite was sought for surviving whales. The whales so hotly pursued in Antarctic waters were the Blue and Finbacks which had been ignored by whalers of the nineteenth century who chased the Sperm, Right, Humpback and Sei whales now no longer abundant. The Grey whale in 1932 was declared a "rarity" and the once important Bowhead of the Arctic regions not a figure in modern whaling at all. "It would be deplorable" wrote Mr. Townsend in *Science* for March 11, 1932 "for last season's slaughter to be repeated in 1933 and the world's most important animal oil source seriously damaged."

Some attempt was at last made at conservation. In September of 1931, the League of Nations drafted an International Convention for the Regulation of Whaling. This Convention between the United States and other powers of the League was concluded at Geneva on September 24, 1931 and ratified by the United States on March 31, 1932. "Despite the small American whaling industry, and non-membership in the League of Nations" advised Raymond M. Gilmore, (Office of Foreign Activities, Department of the Interior, January 25, 1950) "the United States cooperated heartily in many of the League's ventures, one of which was the *regulation* of whaling. The United States Government felt that it could well serve the interests of conservation by participation, because they had no industry and hence no axe to grind. There are pros and cons to this argument. Needless to say, at the behest of the former head of the Carnegie Institution of Washington, Dr. John C. Merriam, the United States has always taken an active interest in international whaling regulations and joined in all treaty provisions. Our delegate to most of these conferences has been Dr. Remington Kellogg, Director, United States National Museum, Washington, D.C."

The regulations of a 1935 agreement prohibited the taking of any Right Blue whale under 60 feet in length, any Fin whale under 50 feet or any female accompanied by a calf. As a conservation measure, the regulations proved a failure. There had been two important omissions. First, no limits were placed on the total catch. Second, the lengths designated were too small. Taking a Blue whale of the permissible 60 foot length meant denial of a chance to reproduce. In 1929 the Discovery Committee of Britain had found that the Blue whale does not become sexually mature until a length of over 70 feet is reached.

Several Conventions have been held in behalf of the whale's salvation. There were various amendments made to improve the 1935 agreement. On August 13, 1937 the Whaling Treaty Act of 1936 was ratified by the President of the United States. All whaling nations subscribed to the terms of this agreement with the exception of Japan. At this period, 1938, the United States had a small whaling fleet of three factory ships, 31 small killer boats, and two shore stations which yielded products valued at $2,000,000. The method of the recalcitrant Japanese was indiscriminately to slaughter everything in their section of the Antarctic. Equal targets were whales on the verge of giving birth and baby whales. This procedure was denounced as "reckless and wasteful" by Dr. Frank Thone of *Science Newsletter* who said that Japanese "disregarded the rights of either the whales or the whalers under other flags, making themselves the most thoroughly unpopular persons in all of the Far South Seas."

With preparations underway for the season in July of 1939, the Norwegians and others who were angered by Japanese non-cooperation and rejection of the joint whaling convention (to which even the Germans had subscribed), openly announced that they were planning a mass attack on whales which would annihilate the whaling population and so "beat the Japanese at their own game," killing the golden-egged goose. The outbreak of war in the autumn of 1939 proved to be the last minute reprieve for the doomed South Polar whales. Some nations withdrew but the Japanese continued to whale, with a fleet of 6 or 7 factory ships and 40 or more killer boats. Pearl Harbor meant a new lease on life to the 10,000 to 12,000 whales the Japanese would have taken in the 1941-42 hunting season.

In February of 1942 nobody seemed to know just where the Japanese fleet was which left for the Antarctic in October of 1941. Of the European whaling fleets, only the English and part of the Norwegian ships were left free to proceed to the Antarctic. The German whalers tied up in Hamburg Harbor were probably blown to bits by the fierce blastings which the RAF bombers unleashed on Hamburg. At least one of the Norwegian ships was captured by the Germans, a few were sunk and others became carriers of gasoline and oil to England. It was Dr. Thone's opinion in 1942 that part of the whaling fleet was at work in the Antarctic because of the pressing need of whale-oil in Britain.

In 1944, an International Agreement was signed by the United Kingdom, United States, Norway, Australia, Canada, South Africa, New Zealand and Eire which defined the conditions under which whaling was to be conducted in the first important post-war season. With the end of the war, big preparations were underway for the resumption of Antarctic whaling which were urged on by a critical world shortage of fats. It was hoped that the International Agreements which proved futile before the war would be enforced and that the whale would win survival for a time.

In the 1946-47 season 30,000 whales were taken. There was no indication that there had been any recovery of the stock during the war years and the percentage of pregnant females had dropped. The last whaling conference at Washington, D.C. in 1946 established an improved International Whaling Convention which was proclaimed on February 5, 1948, following acceptance by Australia, Canada, Denmark, France, Great Britain, Iceland, New Zealand, Norway, South Africa, Soviet Russia and the United States. The Japanese, then under United States administration, had to abide by the Convention laws.

In the 1947-48 season, two more Norwegian factory ships were added and 31,000 whales were taken by all participants. The 17,500 Blue whales killed were considerably in excess of the internationally allowable quota of 16,000 Blue whale units, a *unit* being 100 feet. This was equivalent to 2 Fins, or 6 Seis, with no Blue whale allowably taken under 70 feet, nor a Finner under 55 feet. A sanctuary was also provided for surviving whales. The waters from approximately Cape Horn to the Ross Sea south of 40° South Latitude is the refuge of Baleen whales at all times. Nor can a Humpback whale be taken anywhere south of 40° South Latitude. Factory ships must report details of their catch every week by radio to the Whaling Regulation Authorities in Norway. These authorities decide the date on which whaling of all factory ships must cease during a season which lasts three months unless the quota is reached earlier.

His greatest enemy, of course, is man who covets his valuable carcass. Then there are enemies of his habitat, for instance, the sword fish which can kill a whale by penetrating to its heart and lungs with his spear. But his most dreaded foe is the *Orca Gladiator,* killer whale, ocean outlaw. This 20 to 30 foot killer travels in packs of forty or so and can be found in every ocean. The Orcas, or wolves of the sea, attack even the largest of whales. Frank Bullen's *Cruise of the Cacholot* describes the cruel butchering of a great Bowhead by three killers who tear off the lower jaw of the whale and devour only the tongue of their victims. Almost every cetacean known has some commercial value except the Orca, or Grampus. The killer whale is left entirely alone by man. As the only whale which feeds on warm-blooded animals, he is also isolated from the whale family. This is the cannibal whale which haunts the shores of the Pribilof Islands in the Pacific to feed on fur seals.

Against these killers of the sea, the whale can hold his own, but man is more than a match for Moby Dick. Like the Orca, man has pursued the whale in every ocean, exterminating species after species until he has at last cornered the great mammals at the ends of the earth in the region of the South Pole. Now whaling is an Antarctic industry. In 1903 this region furnished only 2.3% of the world total. As whales decreased elsewhere, more and more ships assembled in the Antarctic which furnished 55% of the world total in 1910, 84% in 1936, and now in 1962, 96% of the world war on whales is prosecuted in the Antarctic whaling theater.

Can the whale manage to survive? There have been many gloomy predictions as to the future of the most valuable of sea quarries. In 1928, authorities were doubtful that the whale supply could stand up under 15,000 killed a year. Nevertheless, the number taken advanced steadily until it reached almost five times that annual figure within the next ten years. There was real danger that Moby Dick would lose his immortality at the hands of relentless and scientific huntsmen.

In 1931, Dr. Martin O. Callaghan held that the whale was probably doomed to early extinction. Some of the species had already been exterminated. Following his destruction in the Gascoygne Bay in the early centuries of his pursuit by the Basques, the hunt was slowly pushed northward to Finland and Iceland and along the western Atlantic. In 1612 the capture of the Greenland whale began in the open waters between Spitzbergen and Greenland, and his pursuit soon extended to David Strait and Baffin Bay. After 200 years of persistent hunting, the whale was driven to the remote places of the Arctic Ocean where he was practically extinct by 1908. With the peaceable Greenland Right whale no longer in the picture, the hunt turned to the much wilder Finbacks. After the interim of half a century, whales were again plentiful in the Antarctic in the early years of the twentieth century.

Hunted from pole to pole, pursued persistently by Northmen, Basques, Americans and Asiatics, the whale has been driven, in turn, from the Mediterranean and Red Seas in the dawn of civilization, from the Bay of Biscay and the western Atlantic in the thirteenth century, from the Bering and Greenland Seas in the nineteenth century, from the Indian and Pacific Oceans in the twentieth century until the ice-ringed Antarctic has become his final refuge.

To conserve the whale the International Whaling Convention intervened. At any rate, the whale has a trump card of his own. When his numbers grow too scarce, his pursuit becomes a losing commercial proposition. Then, with the sky as his canopy and the seas his realm, the rich monarch roams the watery wastes in undisturbed majesty. Twenty-five years ago various replies were given to the query: "How much longer can the world's supply of whales last?" Some whaling men believed they would be extinct for commercial purposes within a decade, while one whaling captain dared to estimate that the whaling population of the North Pacific alone reached a million whales.

This skipper may have gone on record with tongue in cheek and the listener took it without the proverbial grain of salt. After all, the total capture of all species for a period of over 1,000 years has been placed at less than one million.

If the whale should be eradicated like the dodo and the bison, it is questionable that nature could ever duplicate the feat of producing so vast and valuable an animal on this planet. The subject of the whale's survival presents another tantalizing question upon which may hinge the whole future of the whale hunt: Are the determined hunters of our era getting at only the fringes of endless herds in the Antarctic, or are the harpooners really destroying the herds themselves?

Bibliography

Of all human occupations few have aroused more interest in a story-loving public than whaling. Of all human adventures the "gull's way and the whale's way" present the greatest challenge to youth . . . still you can search in vain for whaling material in school texts. You must seek and find the whaling books where the devotees of Old Neptune point the salty way. Some sailing directions for a whaling voyage to which you may add your favorite Atlas or Geography book:

Allen, Glover M. *Whalebone Whales of New England,* Boston, 1916. Mem. Boston Society, Nat. Hist. 8, No. 2, 105.

Anderson, Adam. *Historical and Chronological Deduction of the Origin of Commerce.* Great commercial interests of British Empire. London, 1764.

Anderson, Florence Bennett. *Through the Hawse-Hole (the true story of a Nantucket whaling captain),* 276 pages. The Macmillan Company. 1932. New York.

Andrews, Roy Chapman. *All About Whales,* Random House. 1954, 148 pages.

——— *Whale Hunting with Gun and Camera.*

Ashley, Clifford W. *The Yankee Whaler,* Houghton Mifflin Co. 1926. Illus. with reproductions of Ashley's paintings. Ashley is the first writer to recognize and characterize as a folk art the whalemen's hobby of Scrimshaw, giving it independent recognition in a full chapter. (Cf A. Hyatt Verrill)

Brown, C. Emerson. *Pocket List of Marine Mammals of Eastern Massachusetts.* Salem, Mass. 1913.

Brown, James Templeman. *Whaleman, Vessels, Apparatus and Methods of the Whale Fishery.* In Fishery Industries of the U.S. 2 Vols. Washington, 1887.

Browne, J. Ross. *Etchings of a Whaling Cruise.* Harper & Bros. New York, 1846. 572 pp. Narrative giving viewpoint of a green hand, in many respects the best personal description of a voyage in an American whaler.

Bullen, Frank Thomas (1857-1915). *The Cruise of the Cacholot.* Deep-sea plunderers, a collection of stories of the sea.

Caras, Roger A. *Antarctica (land of frozen time).* Chilton Company, Philadelphia, Pa. 1962.

Chatterton, Edward Keble. Whalers and Whaling, N.Y. Wm. Farquhar Payson. 1931. 251 pp. Good British point of view.

Cheever, Henry Theodore (1814-1897). *The Whale and His Captors (voyage aboard the Commodore Preble).* Published in 1849.

Chippendale, Captain H. A. *Sails and Whales.* Houghton Mifflin Company, Boston, 1951. Widely reviewed, catalogued in libraries and in bibliographies, the late author of this book (for whom Chippendale was a borrowed surname along with the captaincy) aroused doubt among maritime history scholars. A search of the records revealed his claim for a consulship to St. Helena for his father was groundless as well as other claims as to himself and his alleged voyages. Subsequently, on November 28, 1951 the writer was in receipt of a letter from Mr. Ernest S. Dodge, Director of the Peabody Museum in Salem: "You will be interested to know that the "Chippendale" book has been removed from sale by the publishers. Chippendale was, apparently, a complete fake and pulled the wool over their eyes. There was so much objection that the book has been withdrawn."

Church, Albert Cook. *Whale Ships and Whaling.* 179 pages. W. W. Norton & Co. New York.

Cook, John A. *Pursuing the Whale.* Introduction by Allan Forbes. Houghton Mifflin, Boston and New York. 1926.

Crosby, Everett U. *Susan's Teeth and Much About Scrimshaw.* 62 pages, illustrated. Tetaukimmo Press, Nantucket Island, Mass. 1955.

Davis, William M. *Nimrod of the Sea, or the American Whaleman.* London, Sampson Low. 1874.

Dodge, George A. *Whaling Voyage in the Pacific Ocean.* Salem, Mass. 1882. Salem Gazette Press. Pamphlet 30 pages. In Essex Institute.

Dow, George Francis. *Whale Ships and Whaling.* Marine Research Society, Salem, Mass. 1925. 446 pp. A pictorial history which contributes nothing on Salem whaling.

Earle, Walter K. *Scrimshaw: Folk Art of the Whalers.* Whaling Museum Society, Inc. Cold Spring Harbor, New York. 36 pages. 1957.

Edwards, Everett Joshua. *Whale Off! History of American Shore Whaling.* Fred A. Stokes Publishers, New York City, 1932. Written from personal experience of Capt. Edwards.

Felt, Joseph B. *Annals of Salem,* Vol. II, 2nd ed. W. & S. B. Ives, Boston, 1849.

Gilbreth, Frank B. Jr. *Of Whales and Women (One man's view of Nantucket history)* Thomas Y. Crowell, New York, 1956. Tongue-in-cheek account of Nantucket by the co-author of *Cheaper by the Dozen.* Inclusion of the Quaker captain with uninhibited native wife and children in the South Seas plus his inhibited Quaker wife and brood in Nantucket.

Goode, G. B. *Fishery Industries of the U.S.* 2 Vols. Washington, 1887. Chapter by James T. Brown, VII, pp. 218-93 gives most detailed account of methods and appliances.

Hart, Joseph C. *Miriam Coffin* (2 vols. N.Y. 1835 and later editions). Written by a whaleman.

Holman, E. P. *The American Whaleman,* 335 pp. Longmans, Green & Co.

Hopkins, William John. *She blows! And Sparm at that!* Illustrated by Clifford W. Ashley. 361 pages. Houghton Mifflin Company, Boston. 1922.

Hough, Henry Beetle. *Whaling Wives* written in collaboration with Emma Mayhew Whiting. 291 pages. Houghton Mifflin, Boston. 1953. Precious human interest material from old records, unpublished letters and diaries discovered in the ancient houses on Martha's Vineyard reveals the story of the women who went whaling with their captain husbands.

Hunt, Cornelius E. (one of the officers). *The Shenandoah, or the Last Confederate Cruiser,* N.Y., G. W. Carleton & Co. 1867. Tells of the destruction of Yankee whalers in Russian waters in the last struggle on sea between the North and South in the Civil War.

Hyde, Michael *Arctic Whaling Adventures.* Oxford University Press. 126 pages. 1955.

Jenkins, James T. *A History of the Whale Fisheries*. London, H. F. & G. Witherby, 1921. 373 pages. Good early history of whaling, especially on the Basques. Excellent bibliography.

Jenings, Walter W. *Introduction to American Economic History*, N.Y. Thomas Y. Crowell Co. 1928. pp. 389-92 on whaling.

Mackintosh, N. A. *The Natural History of Whalebone Whales*. Published by Discovery Investigations, 52 Queen Anne's Chambers, London, S.W.1, England, 1945. Scientific work in the Antarctic was begun in 1925 focused on natural history of the whale carried on by Royal Research Ship *Discovery*, R.R.S. *Discovery II*, R.R.S. *William Scoresby*, from Marine Station established at South Georgia. Work suspended during World War II, but by 1945 22 volumes of the Discovery Reports were published by the Cambridge Press. N. A. Mackintosh, Director of Research to the *Discovery Committee* in Britain.

Macy, Obed, *History of Nantucket*, 1835. One of the standard histories on whaling.

Macy, William Hussey, *There She Blows*, 1872. Written by a whaleman. Also known as *The Log of the Arethusa*. Lee & Shepard, Boston, 1877.

Melville, Herman. *Moby Dick or the Whale*. American whaling classic giving personal experiences of the author. First published in 1851 in New York.

Moby Dick, The White Whale. Edited by W. Somerset Maugham as one of his ten greatest novels of the world. Introduction by Mr. Maugham. Illustrations by Anton Otto Fisher.

Moby Dick, The White Whale. Illustrations in color by Mead Schaeffer. Large 8 vo.

Moby Dick, the Willard Thorp edition.

Morinière, Noel de la. *Histoire generale des Pèches*, 1815. Vol. I, p. 218, etc. On the Basques as whalemen.

Morison, Samuel Eliot. *The Maritime History of Massachusetts 1783-1860*. Houghton Mifflin Co. 1921. 420 pp. Good chapter on Whaling and excellent Bibliography.

Nichols, George (1778-1865) *Salem Shipmaster and Merchant*. Autobiography. The Salem Press, 1914. Salem, Mass. 89 pp.

Nordhoff, Charles, *Whaling and Fishing*. Cincinnati, 1856. An able seaman serves under an upright whaling captain.

Nordhoff, Charles and James Norman Hall.

Mutiny on the Bounty. Tale of mutiny against Captain Bligh in the South Seas.

Men Against the Sea. The story of Capt. Bligh's 3,600 mile trip in an open boat.

Pitcairn Island. The story of what happened to the mutineers who fled from Tahiti to Pitcairn Island.

The Bounty Trilogy. The Wyeth Edition of the three tales of the *Bounty,* Capt. Bligh and the mutineers in one volume. Illus.

The Hurricane. Devastation of an island in the South Seas.

Botany Bay. The story of the penal settlement that was Australia's first colony in the 18th century.

Osgood, Charles S. and Batchelder, H. M. *An Historical Sketch of Salem,* 1626-1879. Salem, Mass. Essex Institute, 1897.

Reynolds, J. N. *Report on Islands discovered by Whalers in the Pacific Ocean* (1835). Contained in 23rd Congress 2d Session House Exec. Doc. III, No. 105.

Ringold, F. M. (U.S. Consul at Pinta, P.I.). Official report summarized in Hunt's Merch. Mag. XLI, 391 (1859). This deals with evil whaling practices.

Robbins, C. H. *The Gam,* New Bedford, 1899. A group of short stories. Written by whalemen.

Ross, Frank Xavier (1914-) *Flashing Harpoons* (*The story of whales and whaling*). Illustrated by John O'Hara Cosgrove. New York. Crowell, 1958.

Sampson, William. *New York City Courts. Mayor's Court. Is a whale a fish? James Maurice vs. Samuel Judd, December 30, 31, 1818.* S. Van Winkle, New York City, 1819.

Sanderson, Ivan T. *Follow The Whale.* Little, Brown, Boston, 1956.

Scammon, Charles M. *Marine Mammals,* San Francisco, 1874. Marvelous drawings of whales.

Scoresby, William, Jr. *Arctic Regions and Northern Whale Fishery.* 2 Vols. Edinburgh, Scotland, 1823. Born in 1789, Scoresby became one of the greatest whaling captains in history, a distinguished explorer and a Fellow of the Royal Society. Made important Greenland observations.

Spears, John R. *The Story of the New England Whalers.* 1908. Macmillan Co. 420 pp. Contains a good account of the Civil War period including the depredations by Southern raiders.

Stackpole, Edouard A. *The Sea-Hunters* (*the great age of whaling*) *New England Whaling 1635-1835.* J. P. Lippincott Co. New York, 1953.

Scrimshaw at Mystic Seaport. The Marine Historical Association, Inc. Mystic, Connecticut, 1958.

Starbuck, Alexander. *History of the American Whale Fishery. Inception to 1876.* In the U.S. Fish Commission Pt. 4 Report for 1875-6. 779 pp. Includes all known voyages of whaling ports.

Stevenson, Charles H. *Whalebone: Its Production and Utilization.* U.S. Bureau of Fisheries. Document No. 626. Washington Govt. Printing Office, 1907.

Tower, Walter S. *A History of the American Whale Fishery.* Philadelphia, 1907. Publication of University of Penn. No. 20. Bibliography and statistics.

Tripp, William Henry. *There Goes Flukes!* Reynolds Printing Co., New Bedford. 1938. Excellent chapter on Ambergris. Whaling voyage in 1925. Personal experiences of author. Illustrated.

Verrill, A. Hyatt. *The Real Story of the Whaler.* Good popular descriptions and history. Illus. N.Y. D. Appleton & Co. 1916. First to include Scrimshaw as a subject—two and a half pages in his chapter on the "Leisure hours" of whalemen.

Villiers, Alan John. *Whalers of the Midnight Sun.* Charles Scribners, N.Y. 1934.

——— *Whaling in the Frozen South.* Story of the 1923-24 Norwegian Whaling Expedition to the Antarctic. Illus. Bobbs-Merrill Co. Indianapolis. 1925. Author, an Australian, enlisted from Hobart where the *Sir James Clark Ross* recruited, commanded by Capt. Carl A. Larsen who built whaling stations at South Georgia and South Shetlands, Antarctic.

Watson, Arthur C. *The Long Harpoon* (New Bedford), 1929.

Whipple, A. B. C. *Yankee Whalers in the South Seas.* Doubleday, New York, 1954.

Wilkes, Charles, U.S.N. *Narrative of the U.S. Exploring Expedition, 1838-1842* (London 1845). Chap. XII gives list of whaling grounds and discusses objectionable whaling practices.

Additional Sources on Modern Whaling

Birkeland, Knut Bergeson. *The Whalers of Akutan. An Account of Modern Whaling in the Aleutian Islands.* Yale University. New Haven. 1926. 171 pages. Personal observations of life at a modern whaling station. Good description of murderous butchery of cutting-in a whale. Working conditions. Isolation.

Brandt, Karl. *Whale Oil. Economic Analysis.* Stanford University, Calif. Food Research Institute. 1940.

Dakin, William John. *Whalemen Adventurers. Story of Whaling in Australasian Waters and other Southern Seas from the Days of Sail to Modern Times.* Angus & Robertson, Ltd. Sydney, Australia, 1934. 248 pages. Illus. Good Bibliography. Includes summary of work from 1904-1914 at South Georgia of Carl Anton Larsen who died at 64 in 1924 (when he was still whaling). Draft for regulation of whaling at Geneva, 1930, of League of Nations.

Dulles, Foster Rhea, *Lowered Boats, A Chronicle of American Whaling.* N.Y., Harcourt, Brace & Co. 1933.

Hawes, Charles Boardman. *Whaling.* Doubleday. New York, 1924.

Jenkins, Dr. James Travis. *Whales and Modern Whaling.* London. 1932. 22 plates. 234 pages. A plea for conservation of whales. Excellent on natural history in relation to need for protection of species. Discusses international regulation.

McCracken, David R. *Four Months on a Jap Whaler.* Robt. M. McBride & Co. 1948. 227 pages. Illus. 90% of Japan's merchant fleet was sunk in World War II and much of its fishing craft. Gen. Douglas MacArthur authorized two whaling fleets to operate in Antarctic waters 1946-47 to secure anticipated 25,000 tons of whale meat plus $4 millions in whale and liver oil. Sole source of income found to offset U.S. exportations of food to

Japan. Author was a representative of Gen. MacArthur on one of the 2 factory ships. He saw to it that the rules of the Whaling Convention were followed. Good reading on life on board a catcher or killer boat.

Morley, Frank Vigor and Hodgson, J. S. *Whaling North & South.* Century Co. N.Y. 1926. 235 pages. Illus. Adventures of gunners Eduard Skentorp, Lars Andersen. Informal narrative of whaling experiences, particularly good of Antarctic region. Good on ceremony to Neptune in crossing Equator, hazing torments.

Ommanney, F. D. *Below the Roaring Forties. Antarctic Journal.* Longman's Green & Co. London, 1938. 16 illus. Experiences of Royal Research Ship *Discovery II* which sailed Sept., 1929 for Falklands for observation and scientific examination of whales at Marine Biological Station. Exciting account of Little America, frozen Antarctic.

Institutional and Newspaper Sources

Essex Institute Historical Collections, Vol. 75, No. 3, July, 1939. *Brief Account of Whaling Industry in Salem 1820-1860* by Richard Price Northey.

Essex Institute Historical Collections, Vol. 72, July 1936. *The Fur Seal Fishery and Salem* by Edward H. Raymond.

The Old Dartmouth Historical Sketches. Quarterly of the Old Dartmouth Historical Society, New Bedford. Nos. 14, 38, 44 and 45. Valuable whaling material.

Old-Time New England: *Boston Street Lighting in the 18th Century* by Leroy L. Thwing.

Peabody Museum. *The Whaling Industry,* Salem, Mass. 10 pp., 1908. *Special Exhibition of Whaling Pictures* from the Collection of Allan Forbes. 1919. Salem, Mass.

Salem Gazette. Aug. 2, 1836. *Whalers of Salem,* Vessels and Masters.

Salem Register. Oct. 17, 1831. *Success to American Industry.*

———Aug. 1, 1836. *Whaling Business in Salem.*

———April 13, 1837. *Lecture on Whale Fishery* by the Rev. Charles W. Upham.

State Street Trust Company, *Whale Fishery of New England.* Brochure. 1915. Boston, Massachusetts.

The Committee for Whaling Statistics, *International Whaling Statistics* XLV, Oslo, Norway, 1961.

Periodical Literature Sources

America
American Mercury
Antiques
Business Week
Canadian Geographical Journal
Current History
Fortune
Golden Book
Illustrated World
Life Magazine
Literary Digest
National Geographic Magazine
Nature Magazine
Natural History
New Yorker, The
New York Times Magazine
Outlook, The
Popular Mechanics
Popular Science Monthly
Reader's Digest
Russian Review
Scribner's Commentator
Science American
Science Digest
Science
Science Newsletter
Scientific American
Scientific Monthly
Time
Travel
Yale Review

Appendix

MASSACHUSETTS WHALING INDUSTRY AT THREE EARLY PERIODS

Vessels annually fitted out			*Gallons of oil brought in.*	
Northern	*Southern*			
		1771 to 1775		
Fishery	*Fishery*	*Total Tonnage*	*Sperm*	*Whale*
183	121	27,840	1,250,785	272,475
		1787 to 1789		
91	31	10,210	251,370	413,595
		1803 to 1806		
30 (both)		9,360	395,640	677,422

Based on Samuel Eliot Morison *Maritime History of Massachusetts,* p. 396, the data for the first two periods being compiled from those in Pitkin, *Statistical View* (1816 ed.) 78-79; for 1803-06 from the tables in the appendix to W. S. Tower, *History of American Whaling Industry*. "The only Massachusetts ports fitting out whalers betweeen 1803 and 1806 were Nantucket and New Bedford, and the only other American whaling ports were Hudson and Sag Harbor, New York and New London, Conn., each of which fitted out an average of one whaler annually." To these ports should be added Salem, Massachusetts, whose Ship *Rachel* made a sealing and whaling voyage to the Antarctic in 1803.

In Essex County, Salem led with its total of 86 whaling voyages as other ports of Beverly, Gloucester, Lynn and Newburyport joined in the whale hunt.

BEVERLY

Brig, *Benjamin Franklin*	1850-1855	5 voyages
Bark, *Eben Dodge*	1853-1854	2 voyages
Brig, *Escol*	1858-1873	10 voyages
Brig, *Gem*	1849-1853	4 voyages
Bark, *Lady Suffolk*	1853-1859	5 voyages
Bark, *N. D. Chase*	1851-1860	5 voyages
Sch. *Thriver*	1862-1866	5 voyages
		36 voyages

GLOUCESTER

Ship, *Lewis*	1853	1 voyage
Ship, *Mount Wollaston*	1833-1834	2 voyages
Ship, *Polly*	1789-1794	2 voyages
Brig, *Sea Horse*	1788-1789	2 voyages
Sch. *Thorn*	1841	1 voyage
Brig, *Two Friends*	1791	1 voyage
Ship, ———	1793	1 voyage
		10 voyages

LYNN

Ship, *Atlas*	1831-1835	4 voyages
Ship, *Clay*	1833-1836	3 voyages
Ship, *Commander Preble*	1834-1851	10 voyages
Ship, *Louisa*	1831-1838	7 voyages
Ship, *Nahant*	1836	1 voyage
Ship, *Ninus*	1835-1844	6 voyages
Ship, *William Badger*	1845-1849	2 voyages
		33 voyages

NEWBURYPORT

Ship, *Adeline*	1834	1 voyage
Brig, *Chance*	1785	1 voyage
Sch., *Georgia*	1866-1868	2 voyages
Sch., *Hannah Grant*	1867-1869	2 voyages
Sch., *Life Boat*	1867-1868	2 voyages
Ship, *Merrimac*	1833-1842	4 voyages
Ship, *Newburyport*	1834	1 voyage
		13 voyages

SUMMARY

ESSEX COUNTY......	Beverly	36
	Gloucester	10
	Lynn	33
	Newburyport	13
	Salem	86
		178 voyages

Comparable whaling statistics for 1835 to 1844 for Salem, New Bedford, Nantucket, New London. Prepared by Richard Price Northey for the Essex Institute Historical Collections, Vol. LXXV, July 1939.

		No. of Ships	*Sperm Oil*	*Whale-Oil*	*Barrels per ship*
1835	Salem	5	2700	2300	1000
	New Bedford	53	66742	30488	1866
	Nantucket	30	38824	4497	1444
	New London	14	11866	14041	1850
1836	Salem	4	5700	....	1425
	New Bedford	53	39154	38243	1460
	Nantucket	20	35154	2188	1867
	New London	13	3198	18663	1681
1837	Salem	11	3000	4120	647
	New Bedford	53	56831	63683	2274
	Nantucket	25	35056	4569	1585
	New London	18	8469	26774	1959
1838	Salem	4	1000	4750	1438
	New Bedford	92	77600	84100	1752
	Nantucket	16	21730	6200	1745
	New London	24	4400	34000	1600
1839	Salem	8	4630	6670	1412
	New Bedford	75	61695	72843	1793
	Nantucket	18	23140	7550	1670
	New London	22	4500	36690	1395
1840	Salem	6	4330	8120	2075
	New Bedford	82	63465	75415	1696
	Nantucket	25	43330	2275	1824
	New London	23	5145	38320	1887

		No. of Ships	Sperm Oil	Whale-Oil	Barrels per ship
1841	Salem	1	275	1300	1575
	New Bedford	57	54860	49555	1832
	Nantucket	24	39891	3508	1808
	New London	18	4115	27890	1777
1842	Salem	6	7450	120	1261
	New Bedford	63	70909	51112	1957
	Nantucket	16	27654	1519	1823
	New London	21	4013	27719	1485
1843	Salem	2	800	1800	1300
	New Bedford	56	61066	40922	1821
	Nantucket	20	30280	1563	1592
	New London	20	4243	36850	2055
1844	Salem	2	3300	....	1650
	New Bedford	76	54509	102992	2073
	Nantucket	21	31290	4461	1703
	New London	23	2296	9816	1874

Some Comments on Whaling Pictures

From the sixteenth century until the present day whaling has been the subject of artists with varying success. Crude early engravings and paintings were followed by the finished productions of Dutch painters of the seventeenth and eighteenth centuries. French and English artists also made contributions to the pictorial record of the age-old industry. Unfortunately, when whaling was in its full vigor in America during the nineteenth century, there existed no great school of art to preserve the whaleman's triumph. Melville's lasting literary contribution found no counterpart in the field of art. Just as Melville's whaling book remained neglected from 1850 to its re-discovery in the 1920's, whaling, as an art subject, seems to have escaped the attention of the recognized marine artists of the time.

The period of America's sailing ship produced good paintings of harbors, naval and merchant ships. The handsome clipper, especially, excited public attention and proud owners often had their favorites portrayed in oils. The slow-sailing whaler, usually a converted merchantman, lacked the grace and beauty of line possessed by the rapid clippers. The whaler's ends were broad to accommodate hundreds of barrels of oil and to support the heavy try-work furnaces. A whaler laid no claims to grace or beauty, although often aspiring to a fine figurehead at its bow. The nattily uniformed seamen of the merchant ships ridiculed the motley whaling crews and laughed at the plodding whalers. Whaling owners prided themselves on the number of whales caught, the number of barrels of oil imported and the tons of whalebone which lined the whaler bottoms. What the whaler needed was a Jean Francois Millet to glorify her utility.

Charles D. Childs reminds us that no proper pictorial record exists of Nantucket's great whaling era, and that the early run of New England whaling pictures lacks vigor and feeling. The lack was deplored by Herman Melville who said that English and American artists presented "only mechanical outlines of things." Melville praised the French artist Garneray for his realistic whaling pictures which showed the conflict of men with the sea and the whale. Albert Van Best of Rotterdam, Holland, was such an artist as Melville looked for. Van Best (or Beest) appeared in New Bedford not later than 1855, allied himself with William Bradford and later with R. Swain Gifford. To Van Best and Gifford, with the assistance of Benjamin Russell of New Bedford are due the most completely authentic and vigorous whaling scenes produced in America:

1. *Sperm Whaling, The Chase*
2. *Sperm Whaling, The Conflict*
3. *Sperm Whaling, The Capture*

These three well-known lithographs were first published in 1859 by Charles Tobert Co. of New Bedford and later by other companies. Some of Van Best's work is now at the New Bedford Public Library, at the Old Dartmouth Historical Society and Whaling Museum at New Bedford and at the George Walter Vincent Smith Gallery in Springfield, Massachusetts. The largest group of pictures by Benjamin Russell is now at the Massachusetts Institute of Technology, the gift of Allan Forbes of Boston in 1941.

In addition to the work of Van Best and his associates, a Scottish artist, says Mr. Childs, gave us "our most coherent and colorful whaling scenes." These are from the brush of Robert Salmon who came to Boston in 1828 and there painted the whalers which had been outfitted by merchants interested enough and affluent enough to provide for their portraits. Salmon lists five whaling canvasses and panels in his catalogue. The Boston Museum of Fine Arts now owns his *Whale Fishery,* painted for a Mr. Bradle.

The whaling prints issued by Currier and Ives were negligible compared to the 7,000 subjects covered by this firm over a span of three-quarters of a century. My sister, Natalean Cassady, and her husband Noble reproduced in a handsome oil painting a rare scene of Arctic whaling after Currier and Ives especially for this edition of *Whaling and Old Salem,* which appears herein in black and white. The smaller firms of Bufford, Endicott and Prang emphasized whaling a little more, but the subject was overshadowed by dozens of others. Among the rarest and best of America's more outstanding whaling prints are the two well-known aquatints of folio size:

Capturing a Sperm Whale, engraved by J. Hill, 1862, from the painting of William Page, reproduced herein. (In the Allan Forbes Collection.)

A *Shoal of Sperm Whales off the Island of Hawaii,* engraved by Hill and published in 1838, from the canvas of Thomas Birch of Philadelphia, the well-known artist of naval engagements.

One of the most historically interesting of the lithographs by L. Prang and Co. is *A View of the Stone Fleet Sailing from New Bedford* on November 15, 1861, issued in 1862, reproduced herein. The drawing was done by Benjamin Russell of New Bedford, noted especially for his *Right Whaling in Behring Straits and its Varieties,* and his *Sperm Whaling and its Varieties.* These lithographs were produced by J. H. Bufford in 1871. Both depict whalemen attending to their duties on the whaling grounds.

Five lithographs resulted from the Arctic whaling disaster of 1871 in which 34 New England whalers were lost. The pictures were drawn by Benjamin Russell and published by Bufford in 1872. The crushed vessels are identified by name and the rescue of the crews is shown. The late nineteenth century artist W. P. Stubbs produced several oils which depict the Nantucket "camel" floating whaleships over the Island's sand bar. Not to be overlooked among good whaling prints are the two large folio Currier and Ives lithographs taken from paintings by the marine artist Louis Garneray, who went to sea at thirteen.

Many Salem whalers have been preserved in oils and water-colors. This is significant in view of the historical neglect Salem whaling has suffered. The pictorial record is strong evidence that Salem was not so obscure a whaling port as is generally supposed. The marine artist Anton Roux painted some of Salem's whalers in water color such as the Bark *Reaper* done in 1823 in Marseilles, France. Peter Mazzinghi of Leghorn, Tuscany, Italy, did a water-color of the Brig *Malay* in 1833 and Pellegrini painted the Bark *Richard* in 1831 in Marseilles.

The Bark *Emerald* was copied by Mr. M. Macpherson from the original by G. M. White. Macpherson also executed in water-color the Ship *Elizabeth.* This was copied from an original engraved on a whale's tooth, which is at the Peabody Museum. Also represented is the Ship *Sapphire* owned by Stephen C. Phillips. A Macpherson copy is delineated herein of the original oil painting which hangs in the Topsfield home of the late James Duncan Phillips.

Perhaps the most outstanding contribution to Salem's collection of whaling pictures is the depiction of the Bark *Richard* of Salem and the Ship *Julian* of New Bedford actively whaling in the South Pacific in 1837. The painting was done by a local artist, Benjamin F. West (1818 to 1854) and shows a run of Sperm whales being attacked by men in seven whaleboats launched from the *Richard,* the *Julian* and three other unnamed whalers. The *Richard* is shown cutting-in a whale and the *Julian*'s try-works emit the thick, black smoke indicative of boiling blubber. The original of the painting in the Peabody Museum, measures ten feet in length and three feet in width, a fitting tribute to Salem's whaling activity in the South Pacific.

The Yankee Whaler (published in 1926) by Clifford W. Ashley, NRG, contains reproductions of this author-artist's incomparable paintings. A fine representation of Ashley's work are the illustrations he prepared for William John Hopkins' book *She blows! And Sparm at that!* of 1922. The New Bedford Public Library has an excellent collection of the paintings of Clifford Ashley as well as those of Albert Bierstadt and local artists.

One of our best known marine illustrators, Gordon Grant, member of the Nautical Research Guild, excelled in etchings and dry-points. The Associated American Artists (711 Fifth Avenue, New York 22, N.Y.) which possesses illustrated brochures of his works is a good source for Grant's whaling pictures.

Sources:

Charles D. Childs. "Thar she blows" (Some notes on American whaling pictures.") *Antiques,* July 1941. George Francis Dow *Whale Ships and Whaling (Pictorial History of Whaling in Three Centuries*). 1925. Peabody Museum. *Special Exhibition of Whaling Pictures from the collection of Allan Forbes, Esq*. July 1–October 1, 1919.

Melville and Moby Dick or The White Whale

The son of a well-to-do New York merchant, Herman Melville of English and Dutch ancestry (his mother was a Gansevoort, propertied and wealthy), was born at #6 Pearl Street, New York City, on August 1, 1819. When he was only thirteen years old, his father's business collapsed into bankruptcy. Shortly after this catastrophe, his father died and thereafter the Melville family led precarious lives of hardship. Melville tutored and also had a clerk's job in Albany, which to him seemed to lead nowhere. It was therefore with something of desperation that he shipped to sea at eighteen as a cabin boy on a merchantman bound for England.

Upon his return from this initial voyage, he taught school for a while but his thoughts still turned seaward. He now undertook a second voyage in 1841. This was for a whaling cruise on the *Acushnet* out of New Bedford. While he had not anticipated an easy voyage, it proved to be not only one of relentless toil but serving under a cruel whaling captain became after eighteen months unbearable. While the *Acushnet* recruited in the Marquesas Islands, Melville jumped ship. Without any deserting companion, he found himself on the Island of Nukahiva where some friendly cannibals took him captive. Four months spent here in the Taipi Valley gave Melville enough to write and dream about the rest of his life, and he knew the Marquesas cycle had run its course.

Escaping from his hospitable captors, and reluctantly relinquishing the charms of his beauteous native paramour, Melville again shipped on a whaler and soon found himself in Tahiti. Having lingered long enough to explore the way of life and enjoyments of the happy Tahitians, he shipped on a third whaler which brought him finally to Honolulu. Deciding he had had enough of whales and whaling, he enlisted from this seaport in the American Navy as a regular seaman on board the Frigate *United States.* In this manner, he reached Boston in the year 1844, was discharged from the Navy and without loss of time began to write of his incredible experiences in the South Seas and a seaman's life.

His initial work was devoted to life among the cannibals and his liaison with the native girl Fayaway: *Typee: A peep at Polynesian life, or four months of residence in a Valley of the Marquesas* (1846). This tale of love and adventure in an exotic clime was quickly followed by an account—in glowing colors—of his further wanderings in the Pacific: *Omoo: A narrative of adventure in the South Seas* (*1847*). *Typee* and *Omoo* enjoyed an immediate popular success. *Mardi*—a philosophical romance—appeared in 1849. In the same year *Redburn* came off the presses. This was an account of his very first voyage made to Liverpool when he was eighteen. With originality and insight, it told of a young man's experience as a sailor. Melville was not sparing of realism. The book contains more of marine brutality and horror than is found in his book *White-Jacket,* which explores the immensity of the ocean in relation to the abyss of self and the human personality within the framework of life at sea. *White-Jacket or the world in a man-of-war,* published in 1850, proved to be a relentless exposure of life on an American navy vessel. It attracted so much editorial attention, it led finally to the official abolition of flogging in this branch of the federal service.

Typee and *Omoo* laid the foundations for his climactic creation of *Moby Dick.* They can be said to be purely personal experience of adventure in a far place of the globe, told with a flowing freshness, vigor and originality which demanded popular attention. Some readers, of course, were discomfited by Melville's distaste for intruding missionaries and his personal entanglement with a native woman. Others were so fascinated and absorbed by his descriptions of island life among the cannibals, they overlooked the author's contrasting of East and West: his emphasis on the code of the islanders under the shelter of their taboos—and close family ties—and the Westerners with their impulse to strike out into new worlds beyond their past, their quest for a "landlessness" which was partially satisfied by life at sea and intervals spent in the Heavenly Isles.

There is no doubt about it that *Typee* and *Omoo* were the germinal beginning of Melville's allegorical and mystical writings.

In *Mardi* he treated rather satirically and inconclusively with man's cultural life, the eternal verities of life, death, and creativity.

At twenty-eight, already a literary lion, Melville married the lovely Elizabeth Shaw, daughter of a Massachusetts judge. Work proceeded from the year of his marriage, 1847 until 1851 when *Moby Dick* was published. This was his supreme undertaking. To the author's bitter disappointment, his triumphal labor was lukewarmly received. By one of those quirks familiar to the annals of publishing, the superb book remained in literary shadows and was, in fact, a lost novel of the 1850's to the 1920's when it finally became the great literature of the modern period. This was achieved by its re-discovery of Herman Melville by his first modern biographer, Raymond Weaver.

Shortly after the appearance of *Moby Dick,* another book appeared: *Pierre, the Piazza Tales,* and in 1857 *The Confidence Man.* After the Civil War, in 1866: *Battle Pieces and Aspects of the War* and in 1888 *John Marr and Other Sailors.* Actually, however, the peak of Melville's literary career was reached when he was thirty-two, for following the tremendous effort of *Moby Dick,* his need for money, his increasing family responsibilities and burdens practically crippled Melville as a writer. What followed *Moby Dick,* seemed to some critics, to be from another pen.

Just as his intimate friend Nathaniel Hawthorne had held a post in the Salem Custom House—and later a consulship in London, Melville was a customs inspector in New York City for many years, from 1866 to 1885. It was after a lapse of some twenty-two years that Melville completed his prose work *Billy Budd.* Death came in 1891.

Moby Dick as a whaling story grew out of two basic facts: the existence by a freak of nature of a white whale and the destruction of a whale ship by a whale. The mystic allegory grew out of Melville's genius.

"One of our harpooneers told me that he caught once a whale in Spitzbergen that was white all over." *A Voyage to Greenland A.D. 1671.* Harris Coll.

Melville had once read a tale of *Mocha Dick,* a fabulous whale known to the South Sea fishery. He was also familiar with the log of Owen Chace of Nantucket, mate of the ill-fated Whale Ship *Essex,* which had been attacked and sunk by a whale. (*Narrative of the Shipwreck of the Whale Ship Essex of Nantucket which was attacked and destroyed by a large Sperm Whale in the Pacific Ocean in 1821.*)

In what he himself described as "a bold and nervous lofty language," Melville weaves his astonishing narrative of the *Pequod,* whaler, Captain Ahab and his crew in the relentless hunt for the enemy whale. Symbolically, the mammal itself is ageless as Melville relates in his chapter *The Fossil Whale.* He tells of detached broken fossils of pre-adamite whales, fragments of their bones being found at the base of the Alps, in Lombardy, in France, in England, in Scotland and in the Southern United States. He tells of the vast skeleton of an extinct monster (forerunner of the whale) being excavated in 1842 on the plantation of Judge Creagh in Alabama. The English Anatomist Owen declared it was a whale of a departed species and callied it the Zeuglodon in his paper before the London Geological Society.

In that wondrous primordial period, "ere time itself can be said to have begun; for time began with man, I obtain (writes Melville) dim, shuddering glimpses into those Polar eternities; when wedged bastions of ice pressed hard upon what are now the Tropics; and in all the 25,000 miles of this world's circumference, not an inhabitable hand's breadth of land was visible. Then the whole world was the whale's and, king of creation, he left his wake along the present lines of the Andes and the Himalayas. Who can show a pedigree like Leviathan? Ahab's harpoon had shed older blood than the Pharaoh's. Methuselah seems a schoolboy. I look around to shake hands with Shem. I am horror-struck at this antemosaic, unsourced existence of the unspeakable terrors of the whale, which, having been before all time, must needs exist after all time, must needs exist after all humane ages are over."

And so Melville renders emotion and idea in the same context into physical action which after more than a century is just as timely in today's changed world (and which will always be timely) because the theme is valid and timeless, pertaining as it does to man himself in relation to his world, what Melville called the "middle ground between the self and the boundaries of the human condition," and which Quentin Anderson (Columbia University) expresses as "our situation in space and time, our uniqueness and our mortality."

The burden of the book is brought to its high point with Captain Ahab's declaration in *The Quarter Deck* of his intention to find and kill the white whale. This leads to the denouement which brings Captain, vessel and all of the crew—except Ishmael—to their watery deaths. By direct maritime experience and research, Melville encompassed the whole of the whaling enterprise while it was yet in its glory as the most fascinating, challenging, roughest and, at times, the most profitable of American enterprises. A towering narrative of the sea, *Moby Dick* diverts completely from the themes of nineteenth century literature with its dependence for plot on social circumstances. Beyond the seafaring lies an allegory which is still the subject of scholarly discussion. Mysticism and philosophy are read into the quest of Ishmael and the determined hunt of Captain Ahab's pursuit of the great white monster whale who had originally cost him his leg. Named for the King who worshipped Baal, Ahab dies in the final encounter with his enemy, Moby Dick, while Ishmael, the proverbial outcast, lives to relate the whole of the wondrous story. Melville's great art was to unite the hunt for the beast with a profound exploration into the meaning of life itself.

The novel opens with the well known words: "Call me Ishmael. Some years ago—never mind how long precisely—having little or no money in my purse, and nothing particular to interest me on shore, I thought I would sail about a little and see the watery part of the world." Thence we follow the young mariner in the streams above New Bedford down to the port and the wild seas beyond it. The year is 1841. Ishmael enters the Spouter Inn, the favored haunt of the whaling men who throng the streets and wharves of New Bedford, greatest whaling port in the world. The host Peter Coffin asks the lad how much rum he would like. Ishmael orders the full tumbler for a shilling and then joins the sailors in a wild hornpipe. This is his first introduction to the world of whales, whalers, to Mate Stubb and some members of the *Pequod*'s crew and it is also the first glimpse of Ahab, the Captain whom he sees lit by a flash of lightning stalking through the rain on his ivory stump.

Coffin shows Ishmael to his room, explaining that he must share his bed with a harpooner, who is out "selling heads." Ishmael is in bed when the heavily tattooed Queequeg comes in, removes his top hat, tail coat and trousers and hops into bed, smoking a kind of tomahawk pipe. He is as shocked to find Ishmael as Ishmael is by his savage appearance. It is nip and tuck for a minute as to whether Queequeg will not decapitate Ishmael before the story begins. Soon, however, they are good friends and Ishmael drops off to sleep with the thought "Better a sober cannibal than a drunken Christian."

The following day being Sunday, Ishmael attends church. The pulpit is a ship's prow and the parson, Father Mapple hoists himself aboard on a rope ladder which he pulls up after him. During his sermon of fire and brimstone, which is more a sea captain's harangue than a holy scripture rendition, Stubb, Flask, Starbuck and their families are seen.

Ishmael and Queequeg walk down to the wharves together and two part owners of the *Pequod*—Bildad and Peleg—sign them on. Ishmael for 1/300 part of the profit and Queequeg, after he has demonstrated his skill with his harpoon, for 1/60 of the profit. In the early morning mist of the following day, the two meet the half-crazed Elijah who warns them against the Captain of the *Pequod*. Elijah predicts: "At sea one day ye'll smell land where there be no land. And on that day, Ahab will go to his grave, but he will rise again within the hour." As he prophesises Ahab is heard walking down the dock on one good leg and one carved of ivory. It is the top-hatted, frock-coated figure of the tall, remote Ahab, glimpsed through the swirling fog.

The *Pequod* lifts anchor and sails southward through the Atlantic. The captain remains in seclusion. Only the thump of his ivory leg on the deck over the heads of the sleeping fo'c'sle betrays his presence on the ship. Meanwhile, Starbuck, Stubb and Flask run a taut ship, keeping the seamen at holystoning the deck when there is no sail to trim. It is during a routine day that the gaunt figure of Ahab suddenly looms up to the quarterdeck. Even the detached aristocrats of the ship's company, the harpooners Queequeg, Daggo and Tashtego stop their card-playing to watch and wonder.

Starbuck orders them aft. Ahab begins to talk. He rouses them with a series of simple questions to a pitch of frenzy which he climaxes by nailing a Spanish gold ounce to the mast as the reward for the man who first sights Moby Dick, the white whale who has bitten off Ahab's leg and who, we learn, is his real goal of this voyage. After a drink all around, and a great oath of vengeance on Moby Dick, the crew's mess learns more about the history of Moby Dick from the Manxman.

A great hunt for whales ensues and with a goodly capture the try-works swing into action. Captain Ahab shows little interest in the catch but studies maps which show him that Moby Dick will be off New Guinea by the new moon in April and he must get to that spot without delay. His fixation cannot be shaken by Starbuck. The *Pequod* crosses the path of the *Samuel Enderby,* Captain Boomer, just as they encounter a great school of whales. Ahab is excited to learn that Captain Boomer has lost his hand to Moby Dick and that the monster cannot be far away. Now Ahab orders an immediate end to the whale hunt, even cutting loose the dead whales tied up alongside the ship and sets out determinedly after Moby Dick.

Now Starbuck talks with Stubb and Flask about replacing the monomaniac Ahab, but they refuse to betray their captain. The accidental death of a lookout brings ill-luck to the ship as the wind drops and the *Pequod* drifts becalmed aimlessly for days on end. Now Queequeg has a premonition of death, asks the carpenter to build his coffin, and then lapses into silence. Ishmael defends the motionless Queequeg when sadistic sailors begin cutting the latter's tattooed skin. Ishmael is attacked by a knife-wielding Portuguese and is saved by the still silent Queequeg. The struggle ends with a cry of "Thar she blows!" sung out by the lookout. This time it is Moby Dick! It is deep night, still Ahab orders the whaleboats lowered for the attack but they cannot approach the white whale. In the morning Moby Dick moves off and Ahab orders his boats to tow the ship. A wind is raised and the chase is on. Ahab gives the lookout the goldpiece and promises the crew his 10% of the profits when they kill Moby Dick.

At this peak of excitement the *Pequod* meets up with the *Rachel* out of New Bedford whose captain is in great distress as his twelve-year-old son is adrift in a longboat and unable to return to the mother ship. Captain Ahab grimly refuses aid in spite of the other Captain's pleadings and Starbuck's entreaties. Ahab's intensity is rewarded soon by sighting Moby Dick. Four longboats set out for the capture. The White Whale sounds deep and as he rises to the surface he drags a boat under while the men in the other boats try to rescue their drowning shipmates. But Ahab, who takes over the harpoon in the boat of Queequeg screams at the men to keep attacking the whale he has so long pursued to its lair.

The vehement monster aroused to colossal fury, now counterattacks, seizes the captain's boat and easily crushes it in his immense jaws. Now Ahab fights from the sea, plunging his harpoon into the tremendous bulk. Raising himself he climbs on to the whale's back with the aid of one of the many irons already embedded in the white hump. In a frenzy, he keeps methodically probing and stabbing as he seeks the whale's vital vulnerable spot. Too late, he finds himself wound in the ropes attached to the harpoons. He is now a prisoner of the huge whale which dives down deep into the heart of the sea. Moby Dick is victor, and Captain Ahab surrenders in death.

Now Starbuck becomes captain of the *Pequod* and orders the two remaining longboats to continue the battle. The enraged Moby Dick surfaces, revealing to the surviving crew the drowned Ahab still tethered to his back. The boats attack once more, but Moby Dick easily smashes them into matchwood. Then, still berserk, the beast crashes head on into the *Pequod* herself, staving in her sides. The hapless vessel sinks. The struggling whalemen perish with the exception of a single survivor of the entire crew. This is the young Ishmael who saves himself by clinging to the coffin that Queequeg had had made for himself by the ship's carpenters. He is picked up by the *Rachel* to whose captain, Ahab had refused aid to save his son.

"Buoyed up by that coffin, for almost one whole day and night, I floated on a soft and dirge-like main. The unharming sharks, they glided by as if with padlocks on their mouths; the savage sea-hawks sailed with sheathed beaks. On the second day, a sail drew near, nearer, and picked me up at last. It was the devious-cruising *Rachel,* that in her retracing search after her missing children, only found another orphan."

* * *

Critical Studies:

Raymond Weaver, *Herman Melville: Mariner and Mystic.* New York, George H. Doran Company, 1921.

Willard Thorp, *Herman Melville: Representative Selections.* New York, American Book Co.

F. O. Matthiessen, *American Renaissance* (chapter on Melville) New York, Oxford University Press, 1941.

Newton Arvin, *Herman Melville.* New York, William Sloane Associates, 1950, American Men of Letters Series.

Richard Chase, *The American Novel and its tradition* (Chapter on Melville). New York, Doubleday & Co., 1957.

Lewis Mumford

William E. Sedgwick

Charles Oslo

Maxwell Geismar

Index

A

Achilles, 104
Adams, John, 23, 128, 135
Ahab, Captain, 75
Alfred, King, 2
Allen, Augustus, 178
  George H., 178
  John F., 63
  Joseph, 190
American Meat Institute, 52
American Museum of Natural History, 52, 216
American Whaling Company, 64
Anderson, Adam, 2
  Lars, 238
Anne, Queen, 5
Anthony, Abraham, 155, 190, 191
Anton, John, 131
Appleton, John, 32
  Jane, 32
  Nathaniel, 32-36
Archer, Capt. Henry, 76, 157
  Capt. Henry, Jr., 25
  John, 63
Ashley, Clifford, 71

B

Ball, William, 63
Bancroft, Thomas P., 63
Barnard, Captain, 80, 81, 181
  Edwin, 63, 65, 179
Barentz, William, 3, 202
Barét, Jeanne, 142
Barney, Captain, 33, 34
Barrett, Captain George, 190
Barrier Whaling Co., 227
Barston, John, 32
Beecher, Henry Ward, 180
Behring, Vitus, 30
Bennett, Frederick Debell, 177
Bennett, William, 128
Berry, Benj., 128
  William, 128
Bigelow, Capt., 58, 80, 181
Binney, Thomas, 240
Blackstone, 5
Bligh, Capt. William, 27, 32, 134, 141, 143
Bodwich, Richard, 173
Bowditch, Dr. Nathaniel, 23
Bowker, Joel, 63
Boyington, 174, 175
Boynton, Mr., 170, 172
Breed, Josiah, 131
Briggs, Enos, 23
  James B., 62
  Stephen, 131, 174
Bullen, Frank, 134, 245
Burgess, Benj. & Sons, 130
Burman, Abner, 137
Burton, Richard E., 127
Butterfly, Madame, 145
Bougainville, Monsieur de, 141, 143
Bradford, Harry B., 234
Bradstreet, Anne Dudley, 95
Brigham, Dr. Hiram, 151
Brooke, Rupert, 143
Brooks, Capt. N. C., 192
Brown, 65
  Capt., 82
  Albert G., 63
  Benjamin, 63
  Manuel, 136, 137
  Robert, 62, 165, 166, 178
Bryant, Timothy, Jr., 65, 78
Byron, Lord, 143

C

Call, J. R., 128
Calvin, 144

Calfbill, James, 13
California Whaling Co., 53
Callaghan, Dr. Martin O., 245
Cambel, John, 136, 137
Canacker, Thomas Osborn, 154
Carnegie Institution, 241
Carnes, Capt. Jonathan, 23
Cartwright, Capt., 59, 78
  Alexander, 179
Carver, George Washington, 49
Catherine, Empress of Russia, 30
Chadwick, Capt., 65, 97, 99, 108, 131
Charles I, 7, 220
Chase, Capt., 82
  Joshua, 136
Chatterton, E. K., 95, 103
Cheever, 131, 186
  Edward E., 127
  Rev. Henry T., 150
  James W., 62, 63, 65, 76, 80, 82, 83, 127-8, 184-5, 194
Childs, Charles D., 17
Chippendale, 24
Christ, Jesus, 150
Church, Mrs. Charles S., 122
  Captain, 58, 76
  of England, 5
Cleaves, Nathaniel, 128
Cleveland, Miss, 123
Clifton, Elmer, 215
Cloutman, Edward, 90
Coffin, Capt., 65, 78, 81, 175
  Frederick, 190
  George, 81, 184
  James J., 189
  Joshua, 189
Colcord, J. C., 88
Coleman, 65
Collier, John, 154-5
Committee for Whaling Statistics, 229
Conant, Roger, 6
Cook, Capt. James, 6, 27-8, 142, 148-9, 189
  John, 51
  William B., 173
Coombs, Elder M., 151
Cotton, Henry G., 90, 130-1
Cowper, William, 194
Crimblish, Capt., 81, 184
  John, 179
Crowninshield, F. B., 23, 63
Crowell, John, 131
Crusoe, Robinson, 139
Cushing, Isaac, 63

D

Daggett, Isaac C., 154, 155, 157-8
Dana, Richard Henry, 134
Darwin, Charles, 140
Delano, Amasa, 134
Dennis, Thomas, 63
Derby, Elias Hasket, 22-3, 30
Dewing, Capt., 59, 77, 182
  Joseph, 64, 80
Dexter, Elisha, 62
Dexter, Capt. Joseph, 78, 81, 120, 152, 154, 157-60, 162-3, 165-6, 178, 184
Discovery Committee of Britain, 10, 242
Diderot, 143
Disraeli, Benjamin, 138
Doane, Capt. Elisha, 83, 186
  Leander, 90
Dobell, Peter, 202
Dodge, Benjamin, 32
  Benjamin, Jr., 32
  George A., 95-98, 102-3, 105-6, 108, 119, 120, 129, 131-3
  Pickering, 32, 63, 218
Donaldson, Alexander, 63
Dorchester Merchants of England, 6
Dow, George Francis, 152
Dowenes, 160
Downes, Mate, 73
  George W., 119, 154-5, 161, 163
Downing, Thomas, 63
Drexbury, Capt., 175
Drowne, 36
Durant, William, 154
Dutch Whaling Company, 4

E

Eagleston, John H., 58, 62, 194
East India Marine Society, 23
Edds, William, 16
Edwards, David B., 154
  Capt. Everett Joshua 19, 20, 214-16-17
  William, 157
Egleston, Capt., 76
Eldridge, Nicholas, 16
Elizabeth, Queen, 12
Endicott, Capt. Moses, 62, 64
Emmons, Capt., 62, 80
English, Philip, 24
Erie Canal, 57

Essex Institute, 32, 97, 127, 152, 167, 179, 218-19
Essex Register, 56

F

Fairfield, 53
Famous Players Lasky Corp., 215
Fanning, Capt. Edmund, 33, 34, 36
Farnsworth, Ledyard, 154, 160-1
Fayaway, 139, 147
Felt, Joseph, 8, 19, 55
Fish, John, 136
Fish & Wildlife Service, 226
Fisher, Capt. Nehemiah C., 125
  Susan Folger, 123
  Thorpe, 63
  Timothy, 136
Flint, Henry, 136
Flores, Manuel, 154
Fogg, Sylvester P., 63
Folger, Capt., 37
  Burzillai T. Capt., 191
  Elisha Capt., 148
  Mayhew, 32
  Susan, 126
Forest River Lead Co., 219
Forrester, John, 62
  Thomas H., 62
Foster, Stephen C., 195
Foyn, Svend, 210, 221
Franklin, Benjamin, 27, 28
Francis, Capt., 79, 181
  John H., 154
Frost, Capt., 21, 26, 29, 86-7
Furness, William, 90

G

Gardner, Capt. John, 24, 190
  Edmund, 148
Gaugin, Paul, 147
Gauss, Eliza, 152
  Capt. Stephen, 152
Gay, John, 217
Gerrish, George F., 90
Gibson, Horace C., 91
Gilmore, Raymond M., 241
Glover, 36
Goodwin, Capt., 58, 76
  William, 220
Grant, Gen., 207
Grant, George A., 122, 214
  Samuel, 63
Gray, Charles, 131
  William, 22
Greaton, George, 90
Gretney, Edward, 128
Green, Col. E. H. R., 215
  Hetty, 215
Gremer, Samuel T., 90
Grove, Samuel, 62
Gunny, John, 178

H

Hakluyt, Richard, 2
Haley, Lady, 29
Haley, Nelson Cole, 134
Hall, Capt., 79, 82-3, 181, 186
  James Norman, 143
  William, 135-6
  Worthin, 178
Halsey, Admiral William F. Jr., 192
Hamilton, Alexander, 23
Hand, Hiram, 128
Harron, W. M., 173
Hayes, Charles, 154
Harrison, General, 219
Hart, Albert Bushnell, 21
Hathorne, Daniel, 198
Hawthorne, Nathaniel, 23-5, 97, 197-8
Hedge, Capt., 78, 82, 130
  Isaac G., 62, 90, 177
Hepplewhite, 24
Hergesheimer, Joseph, 215
Hersey, Mr., 69, 172-3
Higginson, Major John, 15-18
  Rev. Francis, 15
  Rev. John, 15
  Nathaniel, 15
Hodges, John I., 64, 173, 227
  Joseph, 59, 63, 65, 77, 80, 182
  N. D. C., 179
Hoit, Capt., 57, 76
Holmes, Edward, 128
Holmes, Capt., 199, 206
  Capt. James W., 84-5, 210
  William, 90
Hough, Henry Beetle, 122, 123
Hoxie, Abraham, 178
Hubbard, Oliver, 63
Hull, Capt., 160
Hulling, Hawens, 90-1
Howard, John Jr., 62
Hunt, Cornelius E., 207
Hussey, Capt. Benj. R., 22, 65, 82-4, 185, 208

Christopher, 21
Oliver C., 179

I

Ingalls, Charles, 120, 154, 162-3
Isaiah, 1
Ishamahiel, Edward, 136
International Whaling Convention, 241, 244, 246
International Whaling Statistics, 229

J

Jackson, Benjamin, 63, 213
Capt., 77, 79, 82, 181
James, 175
Joseph, 154, 161
James, Thomas, 173
Jefferson, Thomas, 23, 28
Jemison, Mary, 142
Jenkins, 51
Jenkins, James T., 96
Jenkins, Col. Laurence Waters, 86-7
Jenks, George W., 63
Jernegan, Laura, 122
Jewett, Capt., 78, 81
Jonah, 1, 7, 8
Jones, Francis, 136
John Paul, 28
Johnson, Samuel, 26

K

Kellogg, Dr. Remington, 242
Kempton, Capt., 76
Kidd, Rear Adm. Isaac C., 193
Kimball, James, 70, 91, 155
King, James, 59, 62-3, 65, 77, 79, 181

L

Lafayette, Marquis de, 28
Lake, Abner, 90, 130
Lakeman, Capt., 83, 185
Lakeman, Edward F., 62-3, 178
Lambert, George, 127
Lambrose, Stephen, 136
Lang, Gilman, 90
Langley, I. M., 91
Larrabee, Israel M., 90
Larsen, Capt. C. A., 221-2
Laruck, Johnson, 173
Laurence, Lewis, 128
Layton, F. G., 213
Leach, Samuel, 63
League of Nations, 241
Lewis, James, 135, 137
Ledyard, John, 27-8
Lee, Gen., 207
Lemir, Anton, 137
Lind, Capt., 83, 186
Charles, 178
Lindall, Timothy, 16
Lippen, Amboy, 91
Lombard, Capt., 77
London, Jack, 143
Longfellow, Henry Wadsworth, 108-9, 147, 152
Loper, James, 14, 217
Lord, Nathaniel J., 63
Losell, Antonis, 128
Loti, Pierre, 143
Lovett, Marton R., 173
Lund, Benjamin, 90
Lyman, William G., 90

M

Mackintosh, N. A., 10
MacSwiggan, 97
Macy, Capt., 85, 208
Obed, 14
Thomas, 17
Zacceus, 15
Magellan, 148-9, 190
Malzard, John, 128
Mammon, 196
Manchester, Capt., 82
Manerell, Joseph, 90
Marine Museum of Mystic, 216
Maritime Industries, 53-4, 226
Mary, Queen, 12
Maurois, André, 145
Marshall, Capt., 85, 209
Martin, Robert, 90
Massachusetts Bay Colony, 7, 8, 15
Massachusetts Historical Society, 15
Mather, Cotton, 149
Richard, 7, 8
Maury, Samuel, 154, 156
Mavehal, Edward, 173
Mayhew, Capt., 78
Melville, Herman, 41, 51, 75, 91, 105, 107, 118, 139, 142-3, 147, 198
Merriam, Dr. John C., 242
Messman, George, 43
Metcalf, Capt., 29

Miles, Isaac, 90
Mills, William, 128
Mitchell, Aaron, 191
Mitchell, John, 173
Moore, James, 173
Morrell, Benj., 36
Morison, Prof. Samuel Eliot, 56, 93, 96, 118, 163
Morse & Crocker, 63
Moser, 36
Mystic Marine Museum, 116

Mc

McIntire, Samuel, 23, 24

N

Nantucket Inquirer, 190
Nantucket Mirror, 190
Nantucket Whaling Museum, 122, 214
National Museum, Washington, D.C., 12
Naumkeag Mills, 196, 218
  Steam Cotton Co., 196
Navy Dept., 208
Neal, Capt., 67, 68, 69, 80, 81, 138, 167, 181
  Benjamin, 173
  William S., 177, 179
Netcher, Capt. George, 77
Newcomb, Alexander, 62, 79
New Bedford Whaling Film Corp., 215
  Museum, 44, 214
Nichols, 159, 164-66
  George W., 31, 154
  John, 73
  Sarah Millett, 153
  William Henry, 22, 119, 131, 152-4, 157, 178
Nolan, John, 65
Nordhoff, Charles, 143
Norton, Capt. Thomas H., 121
Nye, Charles, 90

O

Ochter, 1, 2, 221
Office of Foreign Activities, Dept. of Interior, 226, 241
Oliver, John, 128
Olympia Whaling Co., 237
Orne, Capt., 38, 76
Osborn, Ezra, 63
  W. H., 128
Osborne, Dr. Henry, 52
Osgood, 9, 10, 12, 13, 208
  John B., 58, 62-3, 65, 76-83, 181, 184-6
  John C., 65, 84-5

P

Pacific Steam Whaling Co., 43
Paddock, Capt., 76
Padduck, Ichabod, 14-16
Page, Edward, 173
Palfray, William W., 63
Paramount Pictures, 215
Pascal, Blaise, 200
Patch, Ira J., 19
Peabody, Col. Francis, 218-19
  Museum, 23, 86, 213
Peach, Francis, 173
  Pen, 128
Perkins, Aaron, 63
  Augustus N., 173, 178
  Thomas H., 32
Perry, Commodore, 191
  Pilot, 158
Philibert, Monsieur, 143
Phillips, James Duncan, 64
  Stephen C., 5, 6, 60, 65, 73, 78, 80-83, 152, 155, 160, 165, 177, 179, 181, 184, 194
Pierce, John B., 182
Pilgrims, 7
Pinkham, Charles F., 185
Pitman, Capt., 58, 77
Pollard, Capt. George, 12
Polo, Marco, 23, 44
Portugal Jim, 163
Potter, Matthew, 154
Powell, George W., 154, 156, 160
Pribilof, Gerassim, 30
Prior, Capt., 83, 199
Producers Trading Co., 54
Pulsifer, Royal E., 90
Putnam, Charles F., 63
Puritan, 7, 23-4, 33, 144, 148, 198, 220, 226

R

Radcliff, Capt., 79, 80, 128, 177, 182
Ramsdell, Capt., 63, 65, 77, 80, 178, 181

Ravens, Joseph, 120, 154, 162, 163
Ray, Capt. George, 190
Rea, Thomas, 63
  William, 63
Richards, George, 123
Richmond, Capt., 85, 210
Ricketson, Capt., 191
Rideout, Chas., 136
Ritchie, Mr., 175-6
Robertson, S. W. & Co., 219
Robinson, Augustus, 128
  Charles, 128
  George, 128
Rochefoucauld, 146
Rogers, Nathaniel L., 63
Rose, Capt., 82
Roosevelt, Franklin D., 71
Roundy, Stephen, 62
Royal Greenland Trade Dept., 227
Royce, Capt., 188, 191, 194, 202
Russell, Capt., 59, 63, 76-7
  Joseph, 20

S

Sabino, Joseph, 136
Salem & South Danvers Oil Co., 219
Salem Evening News, 219
Salem Gazette, 37, 59, 97
  Register, 181
Saltonstall, Leverett, 64
Sanborn, J., 128
  Theophilus, 63
Sancho, King, 2
Sanderson, Ivan T., 8, 234, 239, 240
Sargent, Moses, 176
Saunders, Jonathan P., 177
Savage, John, 14
Savory, Mr., 165
  Nathaniel, 191
  Richard, 63
Sawyer, Asa, 63
Secomb, Ebenezer, 219
Senate Naval Affairs Committee, 191
Science Digest, 225
Science Magazine, 241
Scoresby, Capt., 39
  William, 10
Semmes, 207
Shakespeare, 39, 61, 184, 206
Shepherd, Mr., 136
Sheraton, 24
Sherman, Christopher, 154
Sherwood, 173
Shirley, Jonathan, 90
Silsbee, Capt., 64
  Nathaniel, 62
  William, 62
  Zachariah F., 62-3
Silver, William, 177
Simonds, Samuel, 62-3
Sistare, Capt., 58, 77-8
Smith, Caleb, 127-29, 218
  John, 6
  Thomas, 16
Slocum, Joshua, 134
Sluman, Thomas, 128
Snelling, John, 131
Societé des Pécheries Coloniales a la Baleine, 227
Sparhawk, Jane, 32
  John, 32
Spears, John R., 2, 18
Stickny, 170
Stickney, William, 63
Stone, Robert, 62
Stull, David C., 45
Stafford, William, 154
Stanton, Wm. R., 178
Starbuck, 14, 20, 41, 58, 65, 76, 87, 182, 186, 212
  Alexander, 61
  Charles, 118
  Capt. Obed, 190
Stevenson, Robert Louis, 143
  Capt., 176
Stoddard, Daniel, 63
  Capt., 33, 34
Sweet, Paul, 135
Swetchine, Madam, 167

T

Teval, Capt., 170
Thomas, Nathaniel, 16
Thomson, Barbara, 142
Thone, Dr. Frank, 242-3
Thurstan, Asa, 151
Tojo, Gen. Hideki, 193
Townsend, Mr., 241
  Ebenezer, 31
  Ebenezer, Jr., 31
Tracy, Capt., 79, 181
Trask, Moses, 90-1
Tripp, William H., 44, 214-15

Tucker, Charles R., 118
  Gideon, 52
Turner, Capt., 191
Tuttle, Henry H., 154, 160-1

U

United States Govt., 241
  National Museum, 242
  Navy, 192-3
United Whales Ltd., 236
Upham, 22
  Elbrige C., 154
  Rev. Charles W., 13-16, 181
Upton, Capt., 76

V

Van Buren, Martin, 184
Verne, Jules, 203
Very, John C. Jr., 193
Villiers, Allan John, 221
Von Bellinghausen, 188

W

Waddell, Lt. James, 207
Wade, Jared, 128
Wallis, Capt. Samuel, 141
Ward, Gen. Frederick Townsend, 215
Ward, James, 179
Waterhouse, Capt., 124
Weatherby, 173
Webb, Benj., 65, 83-4, 199, 208
Webster, Noah, 117, 118
Welch, Patrick, 173
Wellder, Capt., 176
West, Benj., 173
  George, 63
  Nathaniel, 32, 78
Western California Fish Co., 53
Weston, Nathaniel, 59, 65, 81, 184
West Virginia, 193
Wilberforce, William, 55
Wilkens, Capt., 156, 162
Wilkes, Lt., 190
  Charles, 188
Williams, Capt., 80
  Charles, 131
  Henry, 173
  John, 12, 128
  Timothy, 32
Wingate, Benj., 173
Whippey, David, 139
White, S., 76
Whiting, Emma Mayhew, 122
Willoughby, Englishman, 3
Willson, John, 90
Wing, J. & W. R., 204
Winship, Capt. Jonathan, 190
Winslow, Capt., 79
  James W., 62
Winthrop, Gov. John, 7
Woodbury, 16
Woodworth, John, 128
World Film Corp., 215
Worth, Capt. J. T., 58, 65, 77, 85, 210
  Paul, 190
Wright, William G., 91
Wyatt, Josiah, 91
Wyer, Capt., 32, 34, 37

Y

York, James, 193

Index of Vessels

A

Acushnet, 139
Adeline, Ship, 60
Alabama, 207
America, 174, 175-6
Ann Parry, 173, 178
Arbella, 7
Arizona, 191, 193
Atlas, 60

B

Balaena, 148, 190, 236, 237
Baltic, Ship, 97, 99, 106, 108
Barclay, 33-4, 36, 175-6
Beagle, H.M.S., 140
Beaver, 139, 190
Bengal, Ship, 25, 27, 57-60, 63, 76-7, 79, 82, 86, 115, 132, 171, 177, 181, 185, 213
Bertha, Bark, 214
Betsey, 33, 36
Bounty, H.M.S., 27, 32, 141, 143
Bowditch, Ship, 128
Britannia, 21, 38, 75-6, 86
By Chance, Brig, 116

C

California, 193
Catherine, Ship, 25, 57, 58, 60, 76, 86, 87, 186
Cavalier, Bark, 59, 63, 77, 79, 86, 115, 181
Chance, 213
Charles W. Morgan, 121, 134, 204, 214-5
Charles Doggett, 50, 76, 87, 186
Clay, Ship, 58, 60, 76, 87
Clifford Wayne, 119, 161
Commander Preble, 60
Concord, Ship, 32-7
Constitution, 135
Cowper, 123, 125

D

Derby, 40, 45, 79, 86, 127, 181, 186
Discovery, 10
Dolphin, 142

E

Eagle, 80, 86-7, 182
Eclipse, 156, 162
Eliza, 58, 64, 72, 76, 80, 82, 86, 127-8, 175, 177, 184-6, 194, 196
Elizabeth, Ship, 40, 45, 61, 62, 64-6, 73, 82-3, 86, 87, 90-1, 113, 130, 132, 154-55, 177-8, 185-6, 195-8
Eliza Burgess, 131, 153
Emeline, 78-9, 86, 181, 186
Emerald, 45, 58-9, 62-4, 72-3, 76-8, 81-3, 86, 115, 118-20, 129-30, 152-4, 156, 158, 160-6, 177-8, 184-7, 198
Empire Victory, 236
Endeavor, 171
Essex, 12, 22, 104
Equator, 148, 190
Eunice, 31

F

Falcon, 12, 45, 84, 85, 86, 115, 206, 208-10
Florida, 207
Fogo, 168
Franklin, 62, 73, 79, 86, 181
Friendship, 168

G

Ganges, 189, 191
General Knox, 21, 26, 38, 76, 86-7
Glove, 67

H

Henry, 45, 64, 82, 83, 86, 178, 185-6
Hercules, 72, 191
Heroine, 135, 168
Hirondelle, 170
Hope, 30, 73
Huntress, 160
Hudson, 176

I

Independence, 190
Independence 2nd, 191
Isaac Hicks, 219
Izette, 25, 57-8, 76-8, 79, 82, 86-7, 115, 132, 173, 174, 175, 177-8, 181, 185

J

James Maury, 40, 45, 50, 59, 77, 80, 82, 86, 181, 185, 208
Jefferson, Ship, 30
Junior, Whaler, 67

L

Leonados, 176
Louisa, Ship, 60
Lydia, 59, 63, 77, 80, 86, 178, 181

M

Mac, Schooner, 40, 62, 73, 78, 80, 86, 181
Malay, 62, 64, 72, 80-1, 83, 86, 179, 181, 184-6, 198
Margaret, 214
Margaretta, 83-4, 86, 197
Mars, 33-4, 36
Maro, 190
Martha, Ship, 68, 135, 171, 176
Mary and Ellen, 194
Maryland, 193
Mayflower, 7
Merrimac, Ship, 60
Messenger, 84, 86, 115, 199, 208, 212
Minerva, Ship, 21, 27, 32-3, 37, 86, 188
Morrison, Ship, 162-3
Morning Star, Bark, 123
Mount Wollaston, 60, 78, 81-2, 86, 115, 184-5

N

Nancy, Brig, 38, 57, 76, 86
Nantucket, 190
Naples, Ship, 157
Neptune, 30-32
Nevada, 193
Newburyport, Ship, 60
New Jersey, 219
Ninus, Ship, 60

O

Oak, 209
Octava, Barque, 171
Oklahoma, 193

P

Palestine, Bark, 59, 64, 78, 81, 86, 179, 184, 186
Pallas, Bark, 25, 57, 60, 76, 86
Para, 72, 84, 86, 179, 208-9
Persia, Brig, 64
Polly, 60, 76, 87
Popmunnett, 201

R

Rachel, Ship, 21-2, 27, 37-8, 86, 188
Rajah, 23
Reaper, Bark, 58-9, 67-9, 77, 80-1, 87, 115, 138, 167, 179, 181, 184
Resolution, 27
Richard, Bark, 41, 59, 63, 77, 80, 87, 182
Roscoe, Brig, 73

S

Said Bin Sultan, Bark, 85, 87, 210, 212-13
Samuel Wright, 58-9, 77-8, 81, 87, 184, 186
Sapphire, Ship, 62, 64, 73, 78, 81, 87, 127, 131, 135-6, 178, 179, 184, 186
Sarah Williams, 153

Sea Horse, 60
Shenandoah, 207
Shepherdess, Barque, 124
Sophia, Bark, 175
Splendid, Bark, 45
Spray, Sloop, 134
States, 29
Statesman, 78, 81, 83, 87, 184, 186, 187, 198
Superior, 188, 191, 194, 202
Superior, Pilot Boat, 157
Syren, 190

T

Tennessee, 193
Topaz, 52
Transit, 184
Two Friends, 60

U

Ulysses, 237
Ulysses Sherman, 169

V

Valkyrie, Schooner, 45
Vigilance, 158

W

Wade, Barque, 119, 161
Wanderer, 214
Whale, The, 7
William H. Shailer, Bark, 85, 87, 209-10, 212

Index
Geographical and National

A

Admiralties, 189, 191
Africa, 23, 31, 56, 85, 88-9, 115, 186, 222, 225, 228
African Coast, 232
Akutan (East Aleutians), 226
Alaska, 30, 188, 191, 201-2, 224, 226
Albany, New York, 90
Aleutians, 27, 30, 191, 226
Amagansett, L.I., 216-7
America, 7, 20, 22, 24, 28-30, 58, 75, 142, 145, 187, 191-2, 196, 202, 215, 220, 223, 240
American, 6, 8, 20, 22, 26-7, 31, 38, 40, 46-8, 51-2, 55, 59, 64, 71, 96, 111, 116, 149, 180-1, 183, 187-90, 193-4, 196, 201-7, 214-5, 226-7, 237, 246
American Indians, 51, 142
Ampanan Bay, 120, 162
Amsterdam, Africa, 31
Antarctic, 10, 11, 20, 27, 30, 37-8, 46, 132, 188-9, 214, 221, 226-30, 234, 237-8, 240-3, 245-6
Arabia, 88
Arabians, 44
Arabian Sea, 2, 8
Arabs, 28
Arctic, 4, 7, 55-6, 59, 99, 188, 194, 201-3, 205, 207, 211, 223, 228, 231, 241, 246
Arctic Ocean, 3, 191
Argentina, 223
Asia, 44, 188, 191
Asiatics, 246
Atlantic, 16, 26, 29, 35, 45, 57, 79, 80, 83-5, 155, 181-2, 199, 201, 207, 210, 216, 220, 223, 228, 231-2, 246
Austria, 81, 153
Australasian Islands, 38
Australia, 87, 115, 224, 226, 228, 233, 243-4
Azores, 20, 195, 228, 232

B

Baffin Bay, 246
Baker Islands, 158-9
Baltimore, Md., 154, 156
Bangor, Me., 154
Barbary Coasts, 23
Barnstable, Cape Cod, 90
Basques, 1-3, 8, 51, 222, 240, 245-6
Batavia (Java), 22, 73, 88
Bay of Biscay, 2, 246
Bay of Whales, 132
Bayonne, 2, 51
Bengal, 22
Bering Sea, 30, 228, 233, 246
Strait, 188, 191, 202, 205, 207
Beverly, Mass., 19, 60, 164, 202
Biarritz, 51
Blubber Hollow (Salem), 219-20
Blubbertown, 4, 203
Bluff Harbor (New Zealand), 213
Bombay, 22, 169
Bonins (Pacific), 189
Borneo, 186, 192
Boston, Mass., 13-15, 21, 23, 29-31, 46, 55, 58, 61-3, 73, 78, 84-5, 90, 96-7, 128, 130, 153, 157-8, 183, 189, 190-1, 202, 204, 208, 211
Bouvet Island (Antarctic), 241
Brazil, 20, 89, 223, 228
Bridgeport, Ct., 183
Bristol, 182
Britain, 47, 53, 55, 59, 196, 211, 223-5, 241

British, 3, 6, 10, 27, 46, 55, 148, 189-90, 201, 211, 223-4, 227-8, 235-7, 241
British Columbia, 224, 228, 232
Bucksport, Maine, 62, 73
Burbon, 170
Buzzard's Bay (Mass), 7, 18, 56, 215

C

Calcutta, 29, 219
California, 53, 60-1, 194-6, 201, 211, 226-7, 232
Cambridge, Mass., 108
Canada, 243-4
Canadian, 53
Canton, 29-31, 34, 36-7, 89, 163, 189, 191
Cape Ann, 6
Cape Brett (New Zealand), 236
Cape Cod, 2, 6, 9, 13, 19, 20, 22, 56, 90, 172
Cape de Verde, 168
Cape Horn, 14, 16, 23, 25, 28, 33-5, 58-60, 103, 116, 123, 139, 141, 163, 186, 244
Cape Lewen (New Holland), 156
Cape Lopez, Gabon, 227
Cape of Good Hope, 28, 31, 155, 159, 169, 172, 179
Cape Province, 228, 232
Cathay, 24
Charleston, South Carolina, 86, 207
Chicago (Ill.), 46
Chile, 35, 115, 187, 211, 228, 233
Chichi Jima, 191
China, 22, 28-9, 30-1, 56, 88, 157, 189, 191
China Sea, 211
Chinese, 30
Cibourre, 51
Cienfuegos (Cuba), 130, 153
Coffin Group (Pacific), 189
Cold Harbor, 202
Cold Spring (New York), 183
Columbia (N.H.), 90
Congo, 232
Connecticut, 27, 31, 90, 182-3, 189, 191, 203, 213
Cuba, 130, 153, 165
Cuttyhunk Island, 157, 214-5

D

Danvers, 219
Dartmouth (Mass.), 6, 56, 183, 188
David Strait, 246
Davis Straits, 201, 205
Deerfield, Conn., 90
Delagoa Bay, 178
Delaware, 8
Denmark, 223, 231, 244
Diamond Head, 193
Diomede Islands, 205
Doubtful Island Bay (New Holland), 163
Dundee, Scotland, 211
Dunedin, New Zealand, 45
Durban, South Africa, 222
Durham, N.H., 90
Duxbury, Mass., 134

E

East Africa, 89
Eastham, Mass., 16
East Haddon, Ct., 183
East Hampton, 14, 19, 216, 217
East India, 73
East Indies, 15, 22
Ebon Atoll (Marshalls), 190
Edgartown, Mass., 18, 56, 125, 158, 183, 202
Ellice Islands, 190
Eire, 243
England, 1, 3-6, 15, 16, 18, 25, 90, 142, 163, 224-5, 243
English, 2, 6, 9, 35, 53, 55, 93, 125, 157, 202, 216, 219, 220, 235, 243
Erie Canal, 56
Eskimo, 2, 239
Etruscan, 146
Eureka, Calif., 226
Europe, 1, 3, 5, 28, 31, 56, 142, 189, 226
European, 47, 142, 145, 203, 243
Europa Rocks, Mozambique Channel Africa, 186
Everett, Mass., 46

F

Fairhaven, Mass., 56, 89, 129, 135, 188, 202, 215
Falkland Islands, 10, 26, 34, 76, 79, 86, 182, 222, 223
Falmouth, Mass., 188
Fall River, Mass., 182
Fanning Island, 33
Far East, 17, 207
Faroe Islands, 231-2

Fields Landing, Calif., 53, 226-7
Fiji Islands, 1, 86, 89, 139, 149, 186
Fijians, 149
Finland, 245
Ford Island, 193
Fort Dauphin (Madagascar), 67-9
Fort George (East Indies), 15
France, 16, 23, 28, 51, 55, 141, 143, 146, 241, 244
French, 145, 157, 163
Frenchmen, 26
French Congo, 227, 232

G

Gabon, 228, 232
Galapagos Islands, 201, 211
Gardner Island, 189
Gardner's Pinnacle, 190
Gascoygne Bay, 245
Gay Head (Mass.), 157
Geneva, 241
German, 4, 9, 47, 49, 224, 238, 243
Germany, 48, 223, 225
Georgia, 208
Geography Bay, Australia, 87, 186
Gilberts, 190
Gloucester, 60
Grand Jason, Falklands, 33-4
Great Barrier Island (New Zealand), 228
Great Britain, 10, 55, 244
Greeks, 1
Greenland, 3, 4, 7, 51, 201, 203, 211, 223, 231-2, 245-6
Greensport, N.Y., 182
Greenwich, 170
Guadalcanal, 192
Guam, 190, 192
Guinea, 20
Gulf of St. Lawrence, 55

H

Hamburg, Germany, 243
Hawaii, 6, 27, 64, 86, 115, 139-40, 148-51, 186, 190, 194
Hebrews, 1
Helgoland, 1, 221
Hobart Town, Tasmania, Aust., 81, 115
Holland, 223-5
Holme's Hole, Mass., 157, 183
Hong Kong, 192
Honolulu, 115, 148, 190, 211
Hudson Bay, 205
Hudson, New York, 22, 183
Hudson River, New York, 61
Hull, England, 5, 55, 211
Humboldt Bay, 226

I

Iceland, 3, 51, 55, 228, 231-2, 244, 246
Ilu River, Guadacanal, 192
Independence Island, 191
Indian Ocean, 44-5, 58, 64, 76-7, 79, 80-3, 85, 115, 152, 155, 165, 169, 178-9, 181-2, 185, 204, 210-11, 246
India, 3, 22, 56
Indians, 51, 116, 142
Indies, 24
Ipswich (Mass.), 8, 19, 164
Island of Tinian, 37
Island of Timor, 211
Italians, 201
Iwo Jima, 192

J

Japan, 51, 191-3, 223-5, 227-8, 232, 242
Japanese, 2, 189-90, 192, 227-8, 243-4
Java, 73

K

Kamchatka, 187, 202, 228, 233
 Peninsula, 191
 River, 30
Kauai, Hawaii, 149
Kealakekua Bay, 27
Kidd Island, 189
Knocker's Hole (Salem), 220
Kodiak, 191
Korean, 193
Kuril Islands, 193, 228, 233

L

Labrador, 201, 231
Ladrones (Mariannas), 190
Larvik (Norway), 46
Leith (England), 47
Le Mare Straits, 34
Liahona, Tonga, 151
Limerick, Me., 90
Liverpool, 47
London, England, 6, 90, 149
Long Island, N.Y., 7, 8, 56, 157
Loper Island, 190

Louisiana, 23
Lynn, Mass., 60, 154, 174, 216

M

Mackenzie, 205
Madagascar, 62, 68, 83-6, 89, 187, 198, 232
Madeira, 228, 232
Makahs, 51
Malaya, 156, 162, 192
Manchester, Mass., 20-1, 38, 76, 86
Marblehead, 158, 220
Maine, 62, 73, 90, 96
Marion, Mass., 56, 202
Marquesas Islands, 37, 83, 139, 149, 186-7, 211
Marianas, 189-90
Maro Reef, 190
Marshalls, 190
Martha's Vineyard, 18, 56, 122-3
Massachusetts, 7, 8, 13, 15, 17, 20, 55-7, 59, 90-1, 93, 96, 108, 127, 134, 149, 153-4, 174, 177, 182-3, 190-1, 194, 202-3, 216, 220
Masafuero, 36-7
Mattapoisett, Mass., 56, 154-5, 157-8
Mba, Fiji, 139
Medford, 62, 73
Mediterranean, 8, 246
Melanesia, 149
Middleton, 90, 219
Midway, 192
Milford Haven, Eng., 6
Mischaum Pt., New Beford, 214
Mississippi River, 28
Mitchell Island, 191
Mobile, West Indies, 82
Monterey Bay, 201
Monticello, Va., 28
Montevideo, 80, 87
Mormons, 151
Mozambique Channel, 83, 86, 186, 198
Mystic, Conn., 116, 121, 183, 188, 215

N

Nantucket, 6, 13-17, 19, 21-2, 33, 56, 58, 60, 64, 68, 89, 90, 97, 102-3, 111-115, 116, 121-2, 126, 129, 139, 148, 158, 172, 182, 187, 188-190, 202, 209-10, 214, 217
Natal, Africa, 228, 232
Natick, R.I., 154
Nazis, 238
Neah Bay, Wash., 51
Netherlands, 227
New Amsterdam, 8
New Bedford, 18, 20, 22, 33, 44-6, 55-6, 58, 71-3, 83-4, 86, 89-90, 121-2, 125, 128, 155, 182-3, 187, 191, 197, 202-4, 206, 208, 209, 211, 213-6
Newburgh, N.Y., 183
New England, 2, 13, 17-19, 21, 23, 30, 42, 55, 56, 58, 61, 67, 90, 96, 110, 117, 139, 140, 144, 148, 189, 191, 194, 196-7, 201-2, 207, 211, 218, 223
Newfoundland, 3, 228, 231-2
New Guinea, 57, 63, 76, 90, 189, 192
New Hampshire, 173, 183
New Haven, Ct., 30-1
New Holland, 81, 96, 119, 155-6, 161, 163, 170, 174, 176, 185
New Islands, 33
New London, 46, 56, 58, 63, 82, 129, 182, 187, 202, 204, 213
Newburyport, 20, 60, 183
Newfield, Me., 90
Newmarket, N.H., 57, 63, 76
Newport, R.I., 183
Newark, N.J., 183
New York, 8, 9, 19, 22, 29, 33, 43, 46, 52, 55, 90, 96, 157, 162, 182, 183, 189, 191, 209, 217, 226, 237
New Zealand, 45, 81, 170, 213, 224, 228, 233, 236, 243
Niutao, 190
Nonauti Is. (Gilberts), 190
North America, 37
North Cape, 215
North Kingston, N.H., 90
Northmen, 246
North Pacific, 30, 205, 233
North Pole, 203, 223
Norway, 1, 46, 51, 223-5, 227, 231, 241, 243
Norwegian, 1, 2, 49, 50, 211, 220-2, 222-4, 226-8, 235, 237-8, 243
Norwich, Ct., 213
Nova Scotia, 6
Nova Zembla, 3, 203
Nukulaelae, 191
Nurakita (Ellice Is.), 191

O

Oahu, Hawaii, 76, 186, 193
Oceana, 150

Ogasawara Shoto, 189
Okhotsk, 123, 126, 187, 191, 202, 207
Okinawa, 192
Ora Tonga, 186
Oregon, 53
Orkneys, 223
Orient, 28, 30
Oriental, 145
Orleans, Mass., 9

P

Pacific, 17, 25, 28, 30, 40, 43, 45, 48, 51, 57-60, 76-8, 80-3, 85, 113, 115, 122-3, 126, 134, 139-42, 144, 148-50, 151-2, 155, 181, 186, 189-94, 196, 201, 205, 207, 210-11, 228, 233, 245-6
Panama Canal, 58
Papeete, Tahiti, 146
Paraiba, 228
Para, Brazil, 89
Paris, 28
Pearl Harbor, 192-3, 243
Pennsylvania, 50, 154, 202, 206
Permambuco, 80
Persian Gulf, 8
Persians, 1
Peru, 211, 228
Petersham, Mass., 90
Philadelphia, 154, 156
Philippines, 22, 192
Phoenix Group (Pacific), 189
Pigeon Island, 189
Pitcairn Island, 122, 143
Plymouth, 7, 183
Point Areans, California, 53
Polynesian, 142, 144-5, 151, 187
Portugal, 154
Portuguese, 201
Poughkeepsie, 183
Portsmouth, N.H., 63, 77, 128, 173, 183
Port Armstrong (Alaska), 226
Port Hobron (Alaska), 226
Pribilof Islands, 30, 245
Providence, R.I., 183
Provincetown, Mass., 56, 183, 202, 204, 209, 213

Q

Queen Charlottes, 226

R

Red Seas, 246
Richmond, California, 227
Rio Janeiro, (N.B. Spelling in nautical use retained instead of Rio de Janeiro), 58-9, 77, 82, 85, 115
Rhode Island, 18, 182-3, 203
Robbinston, Me., 62, 73
Rochester, Mass., 56, 183
Rockingham, N.H., 90
Ross Sea (Antarctic), 132, 240, 244
Russia, 27-8, 30, 201-2, 224-5
Russian, 30, 125, 188, 202, 207, 228
Russian Far East, 202
Ryuku Islands, 232

S

Sacramento, 195
Sag Harbor, New York, 22, 58, 183, 187-8, 191, 202, 217
Saipan, 190, 192
Salem, Mass., 7-9, 13, 15-6, 20-7, 29-31, 33, 37-8, 40, 45, 56-65, 67, 71-3, 75, 85-91, 95, 97, 115, 119-122, 127-8, 130, 133, 139, 152-8, 164-5, 168, 173, 177-8, 180-1, 182-91, 194-5, 197-8, 202, 208, 213, 216, 218, 220, 222
Samoa, 139, 192
Saint Helena, 115, 157, 172
St. Lawrence, 3, 8, 55
St. Mary's (Chile), 35, 36, 68
St. Paul Island, Africa, 31
Sandwich Islands, 6, 86
San Francisco, Calif., 46, 52-3, 189, 191, 195-6, 201-2, 207
Savannah, Georgia, 208
Sandefjord, Norway, 46
Scandinavians, 93
Scotland, 3, 211
Sebastian, City of, 2
Shetlands, 30, 76, 223
Siberia, 126, 187, 202
Singapore, 162-3
Slava, 227
Smeerenburg, 4, 203
South Africa, 155, 222-3, 225, 243-4
South America, 35, 38, 88, 178, 181, 209
South Atlantic, 25, 57, 64, 76-9, 80, 89, 152, 155
South Carolina, 208

South Island, 189
South Georgia (Antarctic), 18, 27, 188, 222, 238, 241
South Orkneys, 223
South Pacific, 58, 86, 126, 151, 177-8
South Polar, 243, 245
South Sandwich Isles, 223
South Shetlands, 38, 223
South Seas, 141-145, 146-7, 149, 177-8, 243
South Sea Islands, 17, 150
Southern Asia, 44
Southampton, 8, 216
Soviet Russia, 203, 227, 244
Sovietskaya Ukraina, 227
Spain, 1, 228, 232
Spaniards, 3
Spanish, 193
Spitzbergen, 35, 203, 245-6
Staten Island, N.Y., 46
Stonington, Ct., 33, 63, 182, 188-9, 191
Straits of Belle Isle, 55
Sumatra, 22-3, 64, 73, 88

T

Tahiti, 82, 139-141, 185, 187
Tahitian, 146
Talcahuano, Chile, 83, 187
Tankan Bay, 193
Tasmania, 81, 115
Tenaru River, Guadacanal, 192
Timor Island, 211
Tinian, 190
Tonga, 151
Tonsberg, Norway, 46
Tophams Cove, New Holland, 176
Trieste, 153
Trinidad Head, 226
Trondheim, Norway, 238

U

Ukraina, 227
Unalaska Islands, 28
Union (U.S.), 207
United Kingdom, 225, 227, 243
United States, 10, 11, 28, 38, 46-8, 56, 59, 103, 135, 151, 153, 163, 180, 184, 187, 189, 192-3, 209, 224, 226, 236-7, 241-3
U.S.S.R., 227
Utah, 151

V

Valparaiso, 115
Vancouver Island, 28
Victoria, Brit. Columbia, 9
Viking, 1, 72, 223, 238
Virginia, 6

W

Waianae, 193
Wake, 192
Wareham, Mass., 56
Warren, Rhode Island, 182, 188
Warrenton, Oregon, 227
Washington, D.C., 11, 192, 242, 244
Washington Island (S.Pac.), 33
Washington State, 51
Wellington, N.Z., 213
Western Islands, Azores, 20, 195
Westfield, Mass., 42
Westport, Mass., 56, 183, 202
West Coast (U.S.A.), 52-3
West Indies, 55, 62, 82, 87, 198, 204
Wilmington, Del., 85, 183
Wood's Hole, Mass., 56

Y

Yankee, 30, 40, 93, 118, 200, 202-3, 207

Z

Zanzibar, 23, 44, 89
Zulu, 101

About the Author

A native Californian, Frances Diane Robotti is a New Yorker by adoption. Long a devotee of world history, her penchant is New England. During a summer enrollment at Harvard for special history studies, she was determined to learn more about the fascinating history of Salem and its potential for a revival of interest in American origins. Salem has not been neglected by novelists who since Nathaniel Hawthorne's day have found it a fertile field with its dark Puritanism, its eerie witchcraft trials and hangings on Gallows Hill, and the imprint it bears of foreign settlers from the seventeenth century on. Mrs. Robotti, in particular, has pioneered in probing the significant European background of American colonization. Her research in old documents has penetrated the mists of those colonizing failures which preceded the settlement of Salem in 1626—as the chief port of debarkation for the great Puritan migrations out of England.

Her first book *Chronicles of Old Salem: An Epic of New England* (1948) was unique in her use of the form of year-to-year annals once popular with monk historians of the Middle Ages. It found ready acceptance among genealogists, schools, colleges, libraries and general readers.

"Mrs. Robotti has long been a student of the Salem scene," wrote the late Carleton D. Morse in his introduction to the first edition of *Whaling and Old Salem,* "from its ships to its beautiful homes and furniture. Her research has included the vigorous characters who conducted its great maritime enterprises and has brought to light many aspects of Salem life." Part of Mrs. Robotti's method of research was personally to interview the descendants of the master navigator Nathaniel Bowditch, Nathaniel Hawthorne, the Phillips clan and many others who imparted priceless information.

One of America's few distaffer marine historians, specializing in the age of sail, Mrs. Robotti's book on whaling is the only major work on the subject in this century by a woman. In private life she is the wife of the food authority Peter J. Robotti, owner of New York's Le Château Richelieu and co-author with him of the Fountainhead title *Much Depends on Dinner (or the tablecloth game)*. While her husband from time to time world travels to appraise the culinary arts, Mrs. Robotti manages to assist him and at the same time research her own materials. "Foods and wines which reflect on men's social customs," she says, "are often the key to many historical truths. The sea, the quest for its domination, are still very much a part of the space age. I think both subjects are elemental and acceptable companions."